Mathematical Analysis Exercises Series

Alessio Mangoni

University series Vol. 1

©2022 Alessio Mangoni. All rights reserved.

ISBN: 9798841635451

This book belongs to the "University" series:

Vol. I) Mathematical Analysis: Exercises Series - ISBN 9798841635451

Vol. II) Mathematical Analysis: Exercises Integrals - ISBN 9798835906697

Vol. III) Mathematical Analysis: Exercises The Study of Functions - ISBN 9798835409990

Dr. Alessio Mangoni, PhD

Scientist and theoretical particle physicist, researcher on high energy physics and nuclear physics, author of many scientific articles published on international research journals, available at the link:

http://inspirehep.net/author/profile/A.Mangoni.1

https://www.alessiomangoni.it

I edition, July 2022

$$\sum_{n=1}^{\infty} a_n(x)$$

Contents

Contents 5

1 Introduction 11

2 Exercises 15

2.1 Exercise 1   15

2.2 Exercise 2   16

2.3	Exercise 3	16
2.4	Exercise 4	16
2.5	Exercise 5	16
2.6	Exercise 6	17
2.7	Exercise 7	17
2.8	Exercise 8	17
2.9	Exercise 9	17
2.10	Exercise 10	18
2.11	Exercise 11	18
2.12	Exercise 12	18
2.13	Exercise 13	18
2.14	Exercise 14	19
2.15	Exercise 15	19
2.16	Exercise 16	19
2.17	Exercise 17	19
2.18	Exercise 18	20

2.19	Exercise 19	20
2.20	Exercise 20	20
2.21	Exercise 21	20
2.22	Exercise 22	21
2.23	Exercise 23	21
2.24	Exercise 24	21
2.25	Exercise 25	21

3 Results 23

3.1	Exercise 1	24
3.2	Exercise 2	25
3.3	Exercise 3	26
3.4	Exercise 4	26
3.5	Exercise 5	27
3.6	Exercise 6	28
3.7	Exercise 7	28
3.8	Exercise 8	29

3.9	Exercise 9	30
3.10	Exercise 10	30
3.11	Exercise 11	31
3.12	Exercise 12	32
3.13	Exercise 13	32
3.14	Exercise 14	33
3.15	Exercise 15	34
3.16	Exercise 16	34
3.17	Exercise 17	35
3.18	Exercise 18	36
3.19	Exercise 19	36
3.20	Exercise 20	37
3.21	Exercise 21	38
3.22	Exercise 22	38
3.23	Exercise 23	39
3.24	Exercise 24	40

3.25	Exercise 25	40

4 Solutions 43

4.1	Exercise 1	43
4.2	Exercise 2	52
4.3	Exercise 3	57
4.4	Exercise 4	69
4.5	Exercise 5	80
4.6	Exercise 6	86
4.7	Exercise 7	88
4.8	Exercise 8	94
4.9	Exercise 9	107
4.10	Exercise 10	114
4.11	Exercise 11	121
4.12	Exercise 12	127
4.13	Exercise 13	139
4.14	Exercise 14	144

4.15 Exercise 15	153
4.16 Exercise 16	159
4.17 Exercise 17	162
4.18 Exercise 18	166
4.19 Exercise 19	176
4.20 Exercise 20	184
4.21 Exercise 21	189
4.22 Exercise 22	197
4.23 Exercise 23	209
4.24 Exercise 24	213
4.25 Exercise 25	216

$$\sum_{n=1}^{\infty} a_n(x)$$

1. Introduction

The series of functions represents one of the typical topics present in the first mathematics exams at the University, in degree courses such as physics, mathematics or engineering. This book is intended to be the ideal reference learn how to solve the typical exam exercises. We propose and solve original exercises of various types and difficulties, with extensive comments. In the first chapter there are the texts of the exercises, in the second only the results, while the third chapter is dedicated to their detailed and complete solutions.

We suggest to try to solve the exercises, comparing the solutions and carefully follow the relative solution in case of errors or any doubts.

The exercises on the behavior of the series involve the study of the pointwise convergence of series of functions of the form

$$\sum_{n=1}^{\infty} f_n(x),$$

where $f_n(x)$ is a sequence of functions, after having determined the domain of definition, called $\mathcal{D}$. The sum can obviously also start from a number other than 1 (often for reasons related to the existence conditions), but the study of the behavior remains unchanged, in fact the modification of a finite number of terms of a series does not modify its behavior. If a series converges then if you add or remove a finite number of terms you will obtain a new series that will continue to converge, usually, to a different value.

Studying the behavior of a series of functions means determining the sets of the variable x where the series is convergent, divergent or indeterminate, so that the union of these three sets (which are, for obvious reasons, disjoint sets) is precisely the domain of definition of the series. In order to obtain this, various criteria and theorems are used, in particular the most used are those related to series with positive terms, where $f_n(x) > 0$, $\forall x \in \mathcal{D}$, or the Leibniz criterion in the case of alternating series, where the terms alternate in sign and can be written in the following form

$$\sum_{n=1}^{\infty} (-1)^n g_n(x),$$

where $g_n(x)$ is a sequence of functions with positive terms, $g_n(x) > 0$. The criteria for positive term series are essentially five:

- comparison criterion;
- asymptotic comparison criterion;

- ratio criterion;
- asymptotic ratio criterion;
- root criterion.

A fundamental theorem is that relative to the absolute series, i.e. the series where each term is give by the absolute value of the corresponding term of a given series, which guarantees the convergence of the starting series if the absolute series converges (called absolute convergence).

$$\sum_{n=1}^{\infty} a_n(x)$$

2. Exercises

2.1 Exercise 1

Study the behavior of the series of functions

$$\sum_{n=1}^{\infty} \frac{(x^2-4)^n}{n(n+1)}$$

2.2 Exercise 2

Study the behavior of the series of functions

$$\sum_{n=1}^{\infty} \frac{(x+1)^{\sqrt{n}}}{n+x}$$

2.3 Exercise 3

Study the behavior of the series of functions

$$\sum_{n=1}^{\infty} \frac{1+\log n}{n^2+\sqrt{n}} (x^2+x-2)^{n^2}$$

2.4 Exercise 4

Study the behavior of the series of functions

$$\sum_{n=1}^{\infty} \frac{(\log x + x)^n}{n^2+3}$$

2.5 Exercise 5

Study the behavior of the series of functions

$$\sum_{n=2}^{\infty} \frac{1}{n^2 \log n} \left(\frac{x^2-9}{2x^2+3x-2} \right)^{n \log n}$$

2.6 Exercise 6

Study the behavior of the series of functions

$$\sum_{n=2}^{\infty} \frac{x+1+nx}{x^2+n^3}$$

2.7 Exercise 7

Study the behavior of the series of functions

$$\sum_{n=1}^{\infty} \frac{(1-x^2)^{n^2} e^n}{n^2}$$

2.8 Exercise 8

Study the behavior of the series of functions

$$\sum_{n=2}^{\infty} \frac{n\sqrt{n}+1}{n^3 \log n} \left(\frac{x^2+x-12}{x^3-1} \right)^n$$

2.9 Exercise 9

Study the behavior of the series of functions

$$\sum_{n=1}^{\infty} \frac{(2-x^2)^n}{1+n+n^2}$$

2.10 Exercise 10

Study the behavior of the series of functions

$$\sum_{n=2}^{\infty} \frac{(\sqrt{x}-3)^n}{\log n}$$

2.11 Exercise 11

Study the behavior of the series of functions

$$\sum_{n=2}^{\infty} \frac{(x-1)^{n^2} \log n}{n^n}$$

2.12 Exercise 12

Study the behavior of the series of functions

$$\sum_{n=2}^{\infty} \frac{(x^2+\log x)^n}{n^{3/2} \log n}$$

2.13 Exercise 13

Study the behavior of the series of functions

$$\sum_{n=1}^{\infty} \frac{(x\sqrt{x})^{n+1}}{\sqrt{n}\log(n+1)}$$

2.14 Exercise 14

Study the behavior of the series of functions

$$\sum_{n=1}^{\infty} \frac{\arctan n}{n} (x^3-1)^n$$

2.15 Exercise 15

Study the behavior of the series of functions

$$\sum_{n=2}^{\infty} \frac{(1-e^x)^n}{n\sqrt{\log n}}$$

2.16 Exercise 16

Study the behavior of the series of functions

$$\sum_{n=1}^{\infty} \frac{x^{1/n}}{\sqrt{n+x^2+3}}$$

2.17 Exercise 17

Study the behavior of the series of functions

$$\sum_{n=2}^{\infty} \frac{(x^2-2x-3)^{2^n}}{n!e^n}$$

2.18 Exercise 18

Study the behavior of the series of functions

$$\sum_{n=1}^{\infty} \frac{1}{n^3} \left(\frac{x}{\log x}\right)^{(n+2)^2}$$

2.19 Exercise 19

Study the behavior of the series of functions

$$\sum_{n=2}^{\infty} \frac{(4x^2-1)^n}{\log(n!)\arctan(e^n)}$$

2.20 Exercise 20

Study the behavior of the series of functions

$$\sum_{n=2}^{\infty} \frac{(8-x^2)^{n+1/n}}{\sqrt{n^3-1}}$$

2.21 Exercise 21

Study the behavior of the series of functions

$$\sum_{n=1}^{\infty} \frac{(8-x^2)^n}{n^2+3n-1}$$

2.22 Exercise 22

Study the behavior of the series of functions

$$\sum_{n=1}^{\infty} \frac{(x(\log x + 2))^{n^2}}{n \arctan n}$$

2.23 Exercise 23

Study the behavior of the series of functions

$$\sum_{n=1}^{\infty} \left(\frac{n+x}{n}\right)^n \frac{1}{n^2}$$

2.24 Exercise 24

Study the behavior of the series of functions

$$\sum_{n=1}^{\infty} \left(\frac{x+3}{\sqrt{x}-2}\right)^n \frac{\log n}{n!}$$

2.25 Exercise 25

Study the behavior of the series of functions

$$\sum_{n=2}^{\infty} \frac{n^{2n}}{n!} \left((x-5)(x^2-1)\right)^{n^2 \log n}$$

$$\sum_{n=1}^{\infty} a_n(x)$$

3. Results

In this chapter we report only the results of the exercises, i.e. the domain $\mathcal{D}$ and the subsets of the domain of the series where it converges, diverges or is indeterminate, the complete solution is in fact included in the next chapter. The results will be reported according to a table of this type

	x_1		x_2	
IND	C	C	C	D

where, in the various intervals, or near a specific value assumed by the variable (for example x_1, x_2), three symbols are shown: C stands for "convergent", D stands for "divergent" and IND stands for "indeterminate". The example table above can be interpreted as follows: "the series is indeterminate for $x < x_1$, convergent for $x_1 \leq x \leq x_2$ and divergent for $x > x_3$". For a convergent series $\forall x \in \mathbb{R}$ we will write, for example,

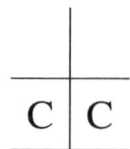

3.1 Exercise 1

Text

Study the behavior of the series of functions

$$\sum_{n=1}^{\infty} \frac{(x^2 - 4)^n}{n(n+1)}$$

Domain

$$\mathcal{D} = \mathbb{R}.$$

Result

	$-\sqrt{5}$		$-\sqrt{3}$		$\sqrt{3}$		$\sqrt{5}$	
D	C	C	C	IND	C	C	C	D

3.2 Exercise 2

Text

Study the behavior of the series of functions

$$\sum_{n=1}^{\infty} \frac{(x+1)^{\sqrt{n}}}{n+x}$$

Domain

$$\mathcal{D} = \{x \in \mathbb{R} : x \geq -1\}.$$

Result

	-1		0	
$\nexists$	C	C	D	D

3.3 Exercise 3

Text

Study the behavior of the series of functions

$$\sum_{n=1}^{\infty} \frac{1+\log n}{n^2+\sqrt{n}} (x^2+x-2)^{n^2}$$

Domain

$$\mathcal{D} = \mathbb{R}.$$

Result

	$\frac{-1-\sqrt{13}}{2}$		$\frac{-1-\sqrt{5}}{2}$		$\frac{-1+\sqrt{5}}{2}$		$\frac{-1+\sqrt{13}}{2}$	
D	C	C	C	IND	C	C	C	D

3.4 Exercise 4

Text

Study the behavior of the series of functions

$$\sum_{n=1}^{\infty} \frac{(\log x + x)^n}{n^2 + 3}$$

Domain

$$\mathcal{D} = \mathbb{R}^+.$$

3.5 Exercise 5

Result

	0		x_0		1	
∄	∄	IND	C	C	C	D

with $x_0 \simeq 0.278$.

3.5 Exercise 5

Text

Study the behavior of the series of functions

$$\sum_{n=2}^{\infty} \frac{1}{n^2 \log n} \left(\frac{x^2 - 9}{2x^2 + 3x - 2} \right)^{n \log n}$$

Domain

$$\mathcal{D} = \{x \in \mathbb{R} : x \leq -3 \vee -2 < x < \frac{1}{2} \vee x \geq 3\}.$$

Result

	-3		-2		$1/2$		3	
C	C	∄	∄	D	∄	∄	C	C

3.6 Exercise 6

Text

Study the behavior of the series of functions

$$\sum_{n=2}^{\infty} \frac{x+1+nx}{x^2+n^3}$$

Domain

$$\mathcal{D} = \mathbb{R}.$$

Result

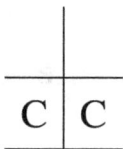

3.7 Exercise 7

Text

Study the behavior of the series of functions

$$\sum_{n=1}^{\infty} \frac{(1-x^2)^{n^2} e^n}{n^2}$$

Domain

$$\mathcal{D} = \mathbb{R}.$$

3.8 Exercise 8

Result

	$-\sqrt{2}$		0		$\sqrt{2}$	
IND	IND	C	D	C	IND	IND

3.8 Exercise 8

Text

Study the behavior of the series of functions

$$\sum_{n=2}^{\infty} \frac{n\sqrt{n}+1}{n^3 \log n} \left(\frac{x^2+x-12}{x^2-1} \right)^n$$

Domain

$$\mathcal{D} = \mathbb{R} - \{\pm 1\}.$$

Result

	$\frac{-1-\sqrt{105}}{4}$		-1	1		$\frac{-1+\sqrt{105}}{4}$		11	
C	C	IND	∄	D	∄	IND	C	C	D	D

3.9 Exercise 9

Text

Study the behavior of the series of functions

$$\sum_{n=1}^{\infty} \frac{(2-x^2)^n}{1+n+n^2}$$

Domain

$$\mathcal{D} = \mathbb{R}.$$

Result

	$-\sqrt{3}$		-1		1		$\sqrt{3}$	
IND	C	C	C	D	C	C	C	IND

3.10 Exercise 10

Text

Study the behavior of the series of functions

$$\sum_{n=2}^{\infty} \frac{(\sqrt{x}-3)^n}{\log n}$$

Domain

$$\mathcal{D} = \{x \in \mathbb{R} : x \geq 0\}.$$

Result

	0		4		16	
$\not\exists$	IND	IND	C	C	D	D

3.11 Exercise 11

Text

Study the behavior of the series of functions

$$\sum_{n=2}^{\infty} \frac{(x-1)^{n^2} \log n}{n^n}$$

Domain

$$\mathcal{D} = \mathbb{R}.$$

Result

	0		1	
IND	C	C	C	D

3.12 Exercise 12

Text

Study the behavior of the series of functions

$$\sum_{n=2}^{\infty} \frac{(x^2 + \log x)^n}{n^{3/2} \log n}$$

Domain

$$\mathcal{D} = \mathbb{R}^+.$$

Result

			0		x_0		1
$\nexists$	$\nexists$	IND	C	C	C	C	D

con $x_0 \simeq 0.330$.

3.13 Exercise 13

Text

Study the behavior of the series of functions

$$\sum_{n=1}^{\infty} \frac{(x\sqrt{x})^{n+1}}{\sqrt{n} \log(n+1)}$$

Domain

$$\mathcal{D} = \{x \in \mathbb{R} : x \geq 0\}.$$

Result

	0		1	
♯	C	C	D	D

3.14 Exercise 14

Text

Study the behavior of the series of functions

$$\sum_{n=1}^{\infty} \frac{\arctan n}{n} (x^3 - 1)^n$$

Domain

$$\mathcal{D} = \mathbb{R}.$$

Result

	0		$\sqrt[3]{2}$	
IND	C	C	D	D

3.15 Exercise 15

Text

Study the behavior of the series of functions

$$\sum_{n=2}^{\infty} \frac{(1-e^x)^n}{n\sqrt{\log n}}$$

Domain

$$\mathcal{D} = \mathbb{R}.$$

Result

	$\log 2$	
C	C	IND

3.16 Exercise 16

Text

Study the behavior of the series of functions

$$\sum_{n=1}^{\infty} \frac{x^{1/n}}{\sqrt{n+x^2+3}}$$

Domain

$$\mathcal{D} = \{x \in \mathbb{R} : x \geq 0\}.$$

Result

	0	
∄	C	D

3.17 Exercise 17

Text

Study the behavior of the series of functions

$$\sum_{n=2}^{\infty} \frac{(x^2 - 2x - 3)^{2^n}}{n!e^n}$$

Domain

$$\mathcal{D} = \mathbb{R}.$$

Result

	$1-\sqrt{5}$		$1-\sqrt{3}$		$1+\sqrt{3}$		$1+\sqrt{5}$	
D	C	C	C	D	C	C	C	D

3.18 Exercise 18

Text

Study the behavior of the series of functions

$$\sum_{n=1}^{\infty} \frac{1}{n^3}\left(\frac{x}{\log x}\right)^{(n+2)^2}$$

Domain

$$\mathcal{D} = \{x \in \mathbb{R} : x > 0 \land x \neq 1\}.$$

Result

	0		x_1		1	
$\nexists$	$\nexists$	C	C	IND	$\nexists$	D

3.19 Exercise 19

Text

Study the behavior of the series of functions

$$\sum_{n=2}^{\infty} \frac{(4x^2 - 1)^n}{\log(n!)\arctan(e^n)}$$

Domain

$$\mathcal{D} = \mathbb{R}.$$

3.20 Exercise 20

Result

	$-\sqrt{2}/2$		$\sqrt{2}/2$	
D	D	C	D	D

3.20 Exercise 20

Text

Study the behavior of the series of functions

$$\sum_{n=2}^{\infty} \frac{(8-x^2)^{n+1/n}}{\sqrt{n^3-1}}$$

Domain

$$\mathcal{D} = \{x \in \mathbb{R} : -2\sqrt{2} \leq x \leq 2\sqrt{2}\}.$$

Result

	$-2\sqrt{2}$		$-\sqrt{7}$		$\sqrt{7}$		$2\sqrt{2}$	
∄	C	C	C	D	C	C	C	∄

3.21 Exercise 21

Text

Study the behavior of the series of functions

$$\sum_{n=1}^{\infty} \frac{(8-x^2)^n}{n^2+3n-1}$$

Domain

$$\mathcal{D} = \mathbb{R}.$$

Result

	-3		$-\sqrt{7}$		$\sqrt{7}$		3	
IND	C	C	C	D	C	C	C	IND

3.22 Exercise 22

Text

Study the behavior of the series of functions

$$\sum_{n=1}^{\infty} \frac{(x(\log x + 2))^{n^2}}{n \arctan n}$$

Domain

$$\mathcal{D} = \mathcal{D} = \mathbb{R}^+.$$

3.23 Exercise 23

Result

	0		x_1	
∄	∄	C	D	D

con $x_1 \simeq 0.642$.

3.23 Exercise 23

Text

Study the behavior of the series of functions

$$\sum_{n=1}^{\infty} \left(\frac{n+x}{n}\right)^n \frac{1}{n^2}$$

Domain

$$\mathcal{D} = \mathbb{R}.$$

Result

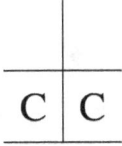

3.24 Exercise 24

Text

Study the behavior of the series of functions

$$\sum_{n=1}^{\infty} \left(\frac{x+3}{\sqrt{x}-2}\right)^n \frac{\log n}{n!}$$

Domain

$$\mathcal{D} = \{x \in \mathbb{R} : x \geq 0 \wedge x \neq 4\}.$$

Result

	0			4	
∄	C	C	∄	C	

3.25 Exercise 25

Text

Study the behavior of the series of functions

$$\sum_{n=2}^{\infty} \frac{n^{2n}}{n!} \left((x-5)(x^2-1)\right)^{n^2 \log n}$$

Domain

$$\mathcal{D} = \{x \in \mathbb{R} : x \leq -1 \ \vee \ x \geq 5\}.$$

3.25 Exercise 25

Result

	$2-\sqrt{10}$		-1		5		$2+\sqrt{10}$	
D	D	C	C	♯	C	C	D	D

$$\sum_{n=1}^{\infty} a_n(x)$$

4. Solutions

4.1 Exercise 1

Text

Study the behavior of the series of functions

$$\sum_{n=1}^{\infty} \frac{(x^2-4)^n}{n(n+1)}$$

Solution

The x variable can take all the values, so the domain is simply $\mathcal{D} = \mathbb{R}$.

We solve the inequality

$$x^2 - 4 > 0.$$

The associated equation $x^2 - 4 = 0$ has the roots $x = \pm 2$, hence

$$x < -2 \quad \vee \quad x > 2.$$

The series is with **positive terms** if $x \in A$ with

$$A \equiv \{x \in \mathbb{R} : x < -2 \vee x > 2\},$$

with **alternate sign terms** if $x \in B$ with

$$B \equiv \{x \in \mathbb{R} : -2 < x < 2\},$$

with null terms, and therefore **convergent**, if

$$x \in Z \equiv \{\pm 2\}.$$

4.1 Exercise 1

We consider the absolute series

$$\sum_{n=1}^{\infty} \left| \frac{(x^2-4)^n}{n(n+1)} \right| = \sum_{n=1}^{\infty} \frac{|x^2-4|^n}{n(n+1)},$$

in fact $\forall n > 1$ we have $n(n+1) > 0$, and we apply the asymptotic ratio criterion

$$\left| \frac{a_{n+1}}{a_n} \right| = \frac{|x^2-4|^{n+1}}{(n+1)(n+2)} \cdot \frac{n(n+1)}{|x^2-4|^n} = |x^2-4| \cdot \frac{n}{n+2},$$

$$\lim_{n \to \infty} \left| \frac{a_{n+1}}{a_n} \right| = |x^2-4| \cdot \lim_{n \to \infty} \left(\frac{n}{n+2} \right) = |x^2-4|.$$

We solve the inequality

$$|x^2-4| < 1,$$

which is equivalent to the system

$$\begin{cases} x^2 - 4 < 1 \\ x^2 - 4 > -1 \end{cases}$$

The first inequality is $x^2 - 4 < 1$, $x^2 - 5 < 0$ whose solution is

$$-\sqrt{5} < x < \sqrt{5},$$

while the second $x^2 - 4 > -1$, $x^2 - 3 > 0$ has the solution

$$x < -\sqrt{3} \lor x > \sqrt{3}.$$

The system becomes

$$\begin{cases} -\sqrt{5} < x < \sqrt{5} \\ x < -\sqrt{3} \lor x > \sqrt{3} \end{cases},$$

whose solution is

$$-\sqrt{5} < x < -\sqrt{3} \lor \sqrt{3} < x < \sqrt{5}.$$

The series is **absolutely convergent** and then also **convergent** if $x \in E$ with

$$E \equiv \{x \in \mathbb{R} : -\sqrt{5} < x < -\sqrt{3} \lor \sqrt{3} < x < \sqrt{5}\},$$

absolutely divergent if $x \in F$ with

$$F \equiv \{x \in \mathbb{R} : x < -\sqrt{5} \lor -\sqrt{3} < x < \sqrt{3} \lor x > \sqrt{5}\},$$

with unknown behavior if

$$x \in S \equiv \{\pm\sqrt{3}, \pm\sqrt{5}\}.$$

4.1 Exercise 1

The values of x for which the series is both absolutely divergent and with positive terms can be obtained intersecting the respective sets A and F.
We obtain

$$D \equiv A \cap F = \{x \in \mathbb{R} : x < -\sqrt{5} \lor x > \sqrt{5}\}.$$

For $x \in D$ the series is **divergent**, because the study of absolute divergence coincides with that of simple divergence, being a series with positive terms.
In the case where the series is absolutely divergent ($x \in F$) and with alternate sign terms ($x \in B$), i.e.

$$x \in G \equiv B \cap F = \{-\sqrt{3} < x < \sqrt{3}\},$$

we can write

$$\sum_{n=1}^{\infty} \frac{(x^2-4)^n}{n(n+1)} = \sum_{n=1}^{\infty} \frac{(-1)^n |x^2-4|^n}{n(n+1)},$$

having used, in this case,

$$x^2 - 4 = -|x^2 - 4|.$$

We can apply the Leibniz criterion, being an alternating series.

Knowing that for $x \in G$ (from absolute divergence)

$$\lim_{n \to \infty} \left| \frac{a_{n+1}}{a_n} \right| > 1,$$

as previously calculated, the positive terms sequence a_n is (from a certain n onwards) not decreasing and therefore the series is **indeterminate** for $x \in G$.

We have to determine the behavior of the initial series for the values $x \in S$ since the asymptotic ratio criterion has not given information, i.e. when

$$|x^2 - 4| = 1.$$

We start with $x = \pm\sqrt{3}$, in this case we have

$$x^2 - 4 = -1,$$

and the series becomes

$$\sum_{n=1}^{\infty} \frac{(-1)^n}{n(n+1)}.$$

4.1 Exercise 1

We apply the Leibniz criterion

$$\lim_{n\to\infty} \frac{1}{n(n+1)} = 0,$$

the positive terms series

$$\frac{1}{n(n+1)}$$

is not increasing, since $n(n+1) = n^2 + n$, in the denominator, is an increasing sequence, and therefore for $x = \pm\sqrt{3}$ the series is **convergent**.

Similarly, for $x = \pm\sqrt{5}$, where $x^2 - 4 = 1$, we have

$$\sum_{n=1}^{\infty} \frac{1}{n(n+1)}.$$

We apply the ratio criterion, we choose $\alpha \in \mathbb{R}^+$ such that

$$\frac{1}{n(n+1)} < \frac{\alpha}{n^2},$$

this happens from a certain n onwards, in this way the series will be convergent, being the generalized harmonic series

$$\sum_{n=1}^{\infty} \frac{1}{n^2}$$

convergent. We solve

$$n^2 < n(n+1)\alpha, \quad \alpha > \frac{n}{n+1} = \frac{1}{1+1/n}.$$

Being

$$\frac{1}{1+1/n} < 1,$$

we can simply choose $\alpha = 1$ and conclude that being

$$\frac{1}{n(n+1)} < \frac{1}{n^2},$$

for $x = \pm\sqrt{5}$, the series is **convergent**.

Summarizing:

The **SERIES** is:

- **CONVERGENT** if

$$-\sqrt{5} \leq x \leq -\sqrt{3} \ \lor \ \sqrt{3} \leq x \leq \sqrt{5};$$

- **DIVERGENT** if

$$x < -\sqrt{5} \ \lor \ x > \sqrt{5};$$

- **INDETERMINATE** if

$$-\sqrt{3} < x < \sqrt{3}.$$

4.1 Exercise 1

Summary scheme:

	$-\sqrt{5}$		$-\sqrt{3}$		$\sqrt{3}$		$\sqrt{5}$	
D	C	C	C	IND	C	C	C	D

4.2 Exercise 2

Text

Study the behavior of the series of functions

$$\sum_{n=1}^{\infty} \frac{(x+1)^{\sqrt{n}}}{n+x}$$

Solution

The base of the power in numerator must be non-negative, because the exponent, i.e. $\sqrt{n}$, assumes non-integer values. It is necessary to write

$$x+1 \geq 0, \quad x \geq -1,$$

the domain is $\mathcal{D} = \{x \in \mathbb{R} : x \geq -1\}$.

For $x = -1$ the series has null terms and then it is **convergent**. We also observe that the series has positive terms from a certain n onwards, in fact the numerator is always positive, while the denominator is positive if $n > -x$. This means that exists $\bar{n} \in \mathbb{N}$ such that for each $n > \bar{n}$ we have

4.2 Exercise 2

$x + n > 0$. We study the following positive term series

$$\sum_{n=\bar{n}}^{\infty} \frac{(x+1)^{\sqrt{n}}}{n+x},$$

in fact, the modification (or cancellation) of a finite number of terms does not modify the behavior of the series. We consider the cases $x > -1$. We use the asymptotic ratio criterion

$$\left|\frac{a_{n+1}}{a_n}\right| = \frac{(x+1)^{\sqrt{n+1}}}{n+x+1} \cdot \frac{n+x}{(x+1)^{\sqrt{n}}},$$

$$\lim_{n \to \infty} \left|\frac{a_{n+1}}{a_n}\right| = \lim_{n \to \infty} (x+1)^{\sqrt{n+1}-\sqrt{n}} \cdot \frac{n+x}{n+x+1}.$$

Knowing that

$$\sqrt{n+1} - \sqrt{n} = (\sqrt{n+1} - \sqrt{n}) \cdot \frac{\sqrt{n+1} + \sqrt{n}}{\sqrt{n+1} + \sqrt{n}}$$

$$= \frac{1}{\sqrt{n+1} + \sqrt{n}}$$

and

$$\lim_{n \to \infty} \frac{n+x}{n+x+1} = 1,$$

we obtain

$$\lim_{n\to\infty}\left|\frac{a_{n+1}}{a_n}\right| = \lim_{n\to\infty}(x+1)^{\frac{1}{\sqrt{n+1}+\sqrt{n}}} = 1$$

then we observe that the criterion does not give additional indications. We use the root criterion

$$\sqrt[n]{a_n} = \left(\frac{(x+1)^{\sqrt{n}}}{n+x}\right)^{1/n} = \frac{(x+1)^{1/\sqrt{n}}}{(n+x)^{1/n}},$$

being

$$(n+x)^{1/n} = e^{\log(n+x)^{1/n}} = e^{\frac{\log(n+x)}{n}} = \exp\left(\frac{\log(n+x)}{n}\right),$$

we have

$$\lim_{n\to\infty}\sqrt[n]{a_n} = \lim_{n\to\infty}\frac{(x+1)^{1/\sqrt{n}}}{\exp\left(\frac{\log(n+x)}{n}\right)} = 1,$$

so even this criterion does not provide indications. We consider the sequence of argument functions of the series,

$$a_n = \frac{(x+1)^{\sqrt{n}}}{n+x}, \quad n > \bar{n},$$

4.2 Exercise 2

we observe that the necessary condition for convergence

$$\lim_{n \to \infty} a_n = 0,$$

is satisfied only if $x+1 < 1$. In fact, we immediately observe that if $x+1 > 1$, i.e. $x > 0$, then

$$\lim_{n \to \infty} \frac{(x+1)^{\sqrt{n}}}{n+x} = \infty,$$

so the series **diverges** (remember that a series with positive terms can only be convergent or divergent). In the case $x = 0$ we have

$$\sum_{n=\bar{n}}^{\infty} \frac{1}{n}.$$

which is **divergent**, being the harmonic series (except for a finite number of initial terms, which do not modify its behavior).

For $-1 < x < 0$, we use the asymptotic ratio criterion ith the generalized harmonic series

$$\sum_{n=1}^{\infty} \frac{1}{n^2}$$

which is convergent, we calculate

$$\lim_{n\to\infty} \frac{(x+1)^{\sqrt{n}}/(n+x)}{1/n^2} = \lim_{n\to\infty} \frac{n^2(x+1)^{\sqrt{n}}}{n+x}$$

$$= \lim_{n\to\infty} \frac{n^2}{(n+x)\left((x+1)^{-1}\right)^{\sqrt{n}}} = 0,$$

being $(x+1)^{-1} > 1$ because $x+1 < 1$. It follows that the series converges thanks to the asymptotic ratio criterion.

Summarizing:

The **SERIES** is:

- **CONVERGENT** if

$$-1 \leq x < 0;$$

- **DIVERGENT** if

$$x \geq 0;$$

Summary scheme:

		-1		0	
♯	C	C	D	D	

4.3 Exercise 3

Text

Study the behavior of the series of functions

$$\sum_{n=1}^{\infty} \frac{1+\log n}{n^2+\sqrt{n}} (x^2+x-2)^{n^2}$$

Solution

There are no restrictions on the values that the x variable can assume, so the domain is $\mathcal{D} = \mathbb{R}$.

We decompose the second degree polynomial. We solve the equation $x^2+x-2=0$, $\Delta = 1+8 = 9$, hence $x_{1,2} = (-1\pm 3)/2$, $x_1 = -2$, $x_2 = 1$, therefore $x^2+x-2 = (x-1)(x+2)$. The series can be written as

$$\sum_{n=1}^{\infty} [(x-1)(x+2)]^{n^2} \cdot \frac{1+\log n}{n^2+\sqrt{n}}$$

We observe that n^2 has the same parity as n, we solve the inequality

$$(x-1)(x+2) > 0,$$

The first factor is positive if $x-1>0$, i.e. $x>1$. The second factor is positive if $x+2>0$, i.e. $x>-2$. Combining the results we get that the product is positive if

$$x<-2 \ \lor \ x>1.$$

So the series is with **positive terms** if $x \in A$ with

$$A \equiv \{x \in \mathbb{R} : x<-2 \lor x>1\},$$

with **alternate sign terms** if $x \in B$ with

$$B \equiv \{x \in \mathbb{R} : -2<x<1\},$$

with null terms, and therefore **convergent**, if

$$x \in Z \equiv \{-2,1\}.$$

We consider the absolute series

$$\sum_{n=1}^{\infty} \left| (x^2+x-2)^{n^2} \cdot \frac{1+\log n}{n^2+\sqrt{n}} \right| = \sum_{n=1}^{\infty} |x^2+x-2|^{n^2} \cdot \frac{1+\log n}{n^2+\sqrt{n}}$$

in fact $\forall n > 1$ we have $n^2+\sqrt{n}>0$ and $1+\log n>0$, and we apply the asymptotic ratio criterion

$$\left| \frac{a_{n+1}}{a_n} \right| = \frac{|x^2+x-2|^{(n+1)^2}}{|x^2+x-2|^{n^2}} \cdot \frac{1+\log(n+1)}{(n+1)^2+\sqrt{n+1}} \cdot \frac{n^2+\sqrt{n}}{1+\log n},$$

4.3 Exercise 3

$$\lim_{n \to \infty} \left| \frac{a_{n+1}}{a_n} \right| = \lim_{n \to \infty} |x^2 + x - 2|^{2n+1}$$

$$= \begin{cases} 0 & \text{if } |x^2 + x - 2| < 1 \\ \infty & \text{if } |x^2 + x - 2| > 1, \\ 1 & \text{if } |x^2 + x - 2| = 1 \end{cases}$$

where we have used

$$\lim_{n \to \infty} \left(\frac{1 + \log(n+1)}{1 + \log n} \cdot \frac{n^2 + \sqrt{n}}{(n+1)^2 + \sqrt{n+1}} \right) = 1.$$

We solve the inequality

$$|x^2 + x - 2| < 1$$

which is equivalent to the system

$$\begin{cases} x^2 + x - 2 < 1 \\ x^2 + x - 2 > -1 \end{cases}$$

The first inequality is $x^2 + x - 2 < 1$, $x^2 + x - 3 < 0$ and has the solution

$$\frac{-1 - \sqrt{13}}{2} < x < \frac{-1 + \sqrt{13}}{2},$$

while the second $x^2+x-2 > -1$, $x^2+x-1 > 0$ has the solution

$$x < \frac{-1-\sqrt{5}}{2} \quad \vee \quad x > \frac{-1+\sqrt{5}}{2}.$$

The system becomes

$$\begin{cases} \frac{-1-\sqrt{13}}{2} < x < \frac{-1+\sqrt{13}}{2} \\ x < \frac{-1-\sqrt{5}}{2} \vee x > \frac{-1+\sqrt{5}}{2} \end{cases},$$

whose solution is

$$\frac{-1-\sqrt{13}}{2} < x < \frac{-1-\sqrt{5}}{2} \quad \vee \quad \frac{-1+\sqrt{5}}{2} < x < \frac{-1+\sqrt{13}}{2}.$$

The series is **absolutely convergent** and then also **convergent** if $x \in E$ with

$$E \equiv \left\{ x \in \mathbb{R} : \frac{-1-\sqrt{13}}{2} < x < \frac{-1-\sqrt{5}}{2} \right. \\ \left. \vee \frac{-1+\sqrt{5}}{2} < x < \frac{-1+\sqrt{13}}{2} \right\},$$

4.3 Exercise 3

absolutely divergent if $x \in F$ with

$$F \equiv \left\{ x \in \mathbb{R} : x < \frac{-1-\sqrt{13}}{2} \right.$$
$$\left. \vee \frac{-1-\sqrt{5}}{2} < x < \frac{-1+\sqrt{5}}{2} \vee x > \frac{-1+\sqrt{13}}{2} \right\},$$

with unknown behavior if

$$x \in S \equiv \left\{ \frac{-1 \pm \sqrt{5}}{2}, \frac{-1 \pm \sqrt{13}}{2} \right\}.$$

We find the values of x for which the series is both absolutely divergent and with positive terms, intersecting the respective sets A and F. We obtain

$$D \equiv A \cap F = \left\{ x \in \mathbb{R} : x < \frac{-1-\sqrt{13}}{2} \vee x > \frac{-1+\sqrt{13}}{2} \right\}.$$

For $x \in D$ the series is **divergent**, because the study of absolute divergence coincides with that of simple divergence, being a series with positive terms.

In the case of an absolutely divergent series ($x \in F$) and with alternate sign terms ($x \in B$), i.e.

$$x \in G \equiv B \cap F = \left\{ \frac{-1-\sqrt{5}}{2} < x < \frac{-1+\sqrt{5}}{2} \right\},$$

we can write

$$\sum_{n=1}^{\infty} ((x-1)(x+2))^{n^2} \cdot \frac{1+\log n}{n^2 + \sqrt{n}} = \sum_{n=1}^{\infty} |(x-1)(x+2)|^{n^2}$$
$$\cdot \frac{1+\log n}{n^2 + \sqrt{n}} \cdot (-1)^n$$

where we used the identity $(-1)^{n^2} = (-1)^n$ and, in this case,

$$(x+1)(x-2) = -|(x-1)(x+2)|.$$

We can apply the Leibniz criterion, being an alternating series. We know that for $x \in G$ (thanks to the absolute divergence)

$$\lim_{n \to \infty} \left| \frac{a_{n+1}}{a_n} \right| > 1,$$

as previously calculated, the positive terms sequence a_n is not decreasing from a certain n onwards, thus the series is **indeterminate** for $x \in G$.

4.3 Exercise 3

Finally we have to obtain the behavior of the initial series for the values $x \in S$ for which the asymptotic ratio criterion has not provided informations, i.e. when

$$|(x-1)(x+2)| = 1.$$

We start with $x = (-1 \pm \sqrt{5})/2$, in this case we have $(x-1)(x+2) = -1$, and the series becomes

$$\sum_{n=1}^{\infty} (-1)^n \cdot \frac{1 + \log n}{n^2 + \sqrt{n}}.$$

We apply the Leibniz criterion, calculating

$$\lim_{n \to \infty} \frac{1 + \log n}{n^2 + \sqrt{n}} = 0.$$

We study the monotony of the positive terms series

$$b_n = \frac{1 + \log n}{n^2 + \sqrt{n}}$$

and we solve, remembering that $n \geq 1$, the inequality

$$b_{n+1} - b_n = \frac{1 + \log(n+1)}{(n+1)^2 + \sqrt{n+1}} - \frac{1 + \log n}{n^2 + \sqrt{n}} < 0,$$

$$\left(1+\log(n+1)\right)(n^2+\sqrt{n}) - \left((n+1)^2+\sqrt{n+1}\right)$$
$$\cdot (1+\log n) < 0$$

$$(2n+1+\sqrt{n+1})(1+\log n) > \sqrt{n}\left(1+\log(n+1)\right)$$
$$+ n^2 \log\left(\frac{n+1}{n}\right)$$

$$\log\left(e^{2n+1+\sqrt{n+1}}\right) + \log\left(n^{2n+1+\sqrt{n+1}}\right) > \log(e^{\sqrt{n}})$$
$$+ \log\left((n+1)^{\sqrt{n}}\right) + \log\left(\left(\frac{n+1}{n}\right)^{n^2}\right)$$

$$\log\left((ne)^{2n+1+\sqrt{n+1}}\right) > \log\left(\left(e(n+1)\right)^{\sqrt{n}}\left(\frac{n+1}{n}\right)^{n^2}\right)$$

$$(ne)^{2n+1+\sqrt{n+1}} > \left(e(n+1)\right)^{\sqrt{n}}\left(1+\frac{1}{n}\right)^{n^2}$$

4.3 Exercise 3

We can write the following relation for the second member of the inequality

$$\left(e(n+1)\right)^{\sqrt{n}}\left(1+\frac{1}{n}\right)^{n^2} < \left(e(n+n)\right)^{\sqrt{n}}\left(\left(1+\frac{1}{n}\right)^{n}\right)^{n}$$

$$< (ne)^{\sqrt{n}} 2^{\sqrt{n}} e^n$$

$$< (ne)^{\sqrt{n+1}} e^{\sqrt{n}} e^n$$

$$< (ne)^{\sqrt{n+1}} (ne)^{n+\sqrt{n}}$$

$$< (ne)^{\sqrt{n+1}+n+\sqrt{n}}$$

and, comparing with the first member of the previous inequality, we solve

$$(ne)^{2n+1+\sqrt{n+1}} > (ne)^{\sqrt{n+1}+n+\sqrt{n}}, \quad (ne)^{2n+1} > (ne)^{n+\sqrt{n}},$$

$$(ne)^{n+1} > (ne)^{\sqrt{n}}, \quad \sqrt{n} < n+1, \quad n < n^2 + 1 + 2n,$$

$$n^2 + n + 1 > 0,$$

true $\forall n \in \mathbb{N}^+$ (in fact $\Delta = 1 - 4 = -3 < 0$).

We have demonstrated the following chain of inequalities

$$(ne)^{2n+1+\sqrt{n+1}} > (ne)^{\sqrt{n+1}+n+\sqrt{n}} > \left(e(n+1)\right)^{\sqrt{n}}\left(1+\frac{1}{n}\right)^{n^2}$$

or better that $b_{n+1} - b_n < 0$, hence the sequence with positive terms b_n is decreasing (and in particular not increasing) $\forall n > 1$.

We conclude that the sequence b_n is not increasing so, for

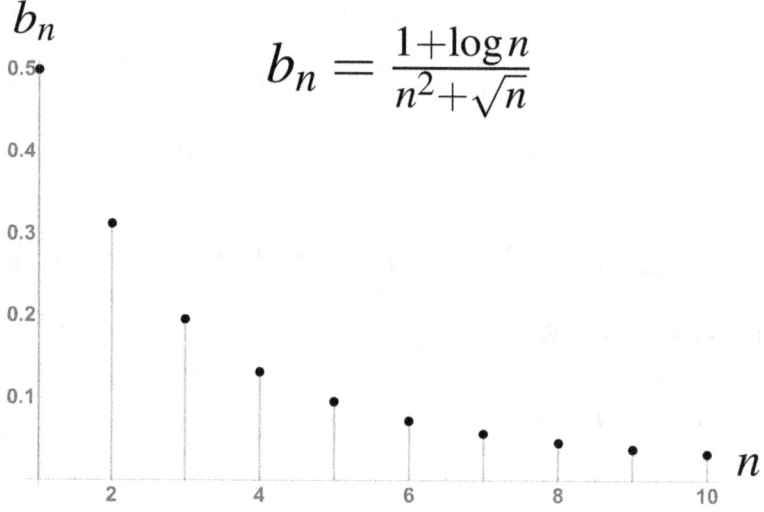

Figure 4.1: Plot of the sequence b_n for $n \in [1, 10] \subset \mathbb{R}$.

$x = (-1 \pm \sqrt{5})/2$ the series is **convergent**. In Figure 4.11 we show the plot for the sequence b_n.

Similarly, for $x = (-1 \pm \sqrt{13})/2$, where $(x-1)(x+2) =$

4.3 Exercise 3

1, the initial series becomes

$$\sum_{n=1}^{\infty} b_n \equiv \sum_{n=1}^{\infty} \frac{1+\log n}{n^2 + \sqrt{n}}.$$

We apply the asymptotic ratio criterion, taking into account the convergent series

$$\sum_{n=1}^{\infty} c_n \equiv \sum_{n=1}^{\infty} \frac{1}{n^{3/2}} = \sum_{n=1}^{\infty} \frac{1}{n\sqrt{n}}.$$

We remember in fact that the generalized harmonic series

$$\sum_{n=1}^{\infty} \frac{1}{n^\alpha}$$

converges if $\alpha > 1$ and diverges if $\alpha \leq 1$.

We calculate the limit

$$\lim_{n \to \infty} \frac{b_n}{c_n} = \lim_{n \to \infty} \frac{n\sqrt{n}(1+\log n)}{n^2+\sqrt{n}} = \lim_{n \to \infty} \frac{\log n}{\sqrt{n}} = 0,$$

concluding that the series, for $x = (-1 \pm \sqrt{13})/2$, is **convergent**.

Summarizing:

The **SERIES** is:

- **CONVERGENT** if

$$\frac{-1-\sqrt{13}}{2} \leq x \leq \frac{-1-\sqrt{5}}{2} \ \lor \ \frac{-1+\sqrt{5}}{2} \leq x \leq \frac{-1+\sqrt{13}}{2};$$

- **DIVERGENT** if

$$x < \frac{-1-\sqrt{13}}{2} \ \lor \ x > \frac{-1+\sqrt{13}}{2};$$

- **INDETERMINATE** if

$$\frac{-1-\sqrt{5}}{2} < x < \frac{-1+\sqrt{5}}{2}.$$

Summary scheme:

	$\frac{-1-\sqrt{13}}{2}$		$\frac{-1-\sqrt{5}}{2}$		$\frac{-1+\sqrt{5}}{2}$		$\frac{-1+\sqrt{13}}{2}$	
D	C	C	C	IND	C	C	C	D

4.4 Exercise 4

Text

Study the behavior of the series of functions

$$\sum_{n=1}^{\infty} \frac{(\log x + x)^n}{n^2 + 3}$$

Solution

To determine the domain we must pay attention to the values that the variable x can assume, in this case we put $x > 0$, due to the presence of the logarithm, so the domain is $\mathcal{D} = \mathbb{R}^+$.

We solve the inequality

$$\log x + x > 0.$$

This inequality cannot be solved analytically (without calculators) so we must proceed graphically, writing $\log x > -x$ and drawing the plot of the function $y = \log x$ and that of the function $y = -x$, as shown in Figure 4.2.

We observe that the two plots intersect in a single point

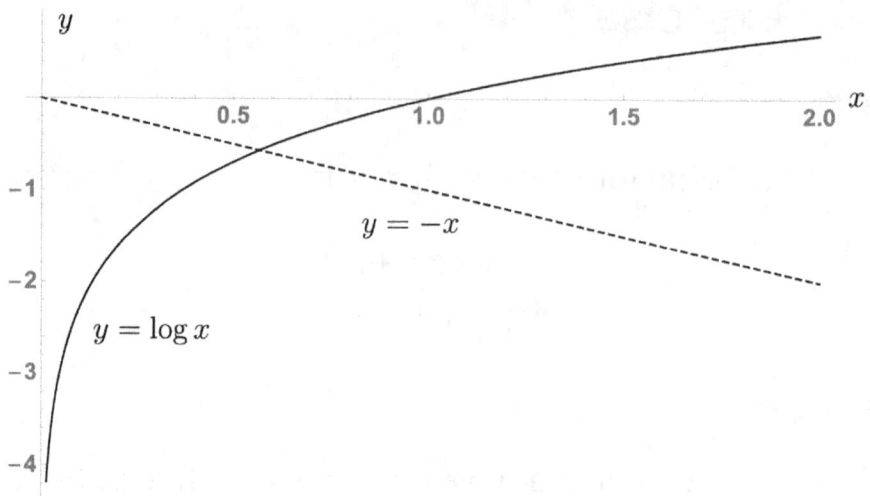

Figure 4.2: Plot of the functions $y = \log x$ (continuous line) and $y = -x$ (dashed line).

with abscissa x_1 between 0 and 1 ($0 < x_1 < 1$). The inequality is satisfied for $x > x_1$.

So the series is with **positive terms** if $x \in A$ with

$$A \equiv \{x \in \mathbb{R} : x > x_1\};$$

with **alternate sign terms** if $x \in B$ with

$$B \equiv \{x \in \mathbb{R} : 0 < x < x_1\};$$

4.4 Exercise 4

with null terms, and therefore **convergent**, if

$$x = x_1.$$

We consider the absolute series

$$\sum_{n=1}^{\infty} \left| \frac{(\log x + x)^n}{n^2 + 3} \right| = \sum_{n=1}^{\infty} \frac{|\log x + x|^n}{n^2 + 3},$$

in fact $\forall n > 1$ we have $n^2 + 3 > 0$, and we apply the asymptotic ratio criterion

$$\left| \frac{a_{n+1}}{a_n} \right| = \frac{|\log x + x|^{n+1}}{(n+1)^2 + 3} \cdot \frac{n^2 + 3}{|\log x + x|^n},$$

$$\lim_{n \to \infty} \left| \frac{a_{n+1}}{a_n} \right| = |\log x + x| \cdot \lim_{n \to \infty} \left(\frac{n^2 + 3}{n^2 + 2n + 4} \right) = |\log x + x|.$$

We solve the inequality

$$|\log x + x| < 1,$$

which is equivalent to the system

$$\begin{cases} \log x + x < 1 \\ \log x + x > -1 \end{cases}.$$

The first inequality can be written as $\log x < 1 - x$ and can be solved graphically. The plots of the functions $y = \log x$ and $y = 1 - x$ are shown in Figure 4.3.

The solution of the associated equation $\log x = 1 - x$ is

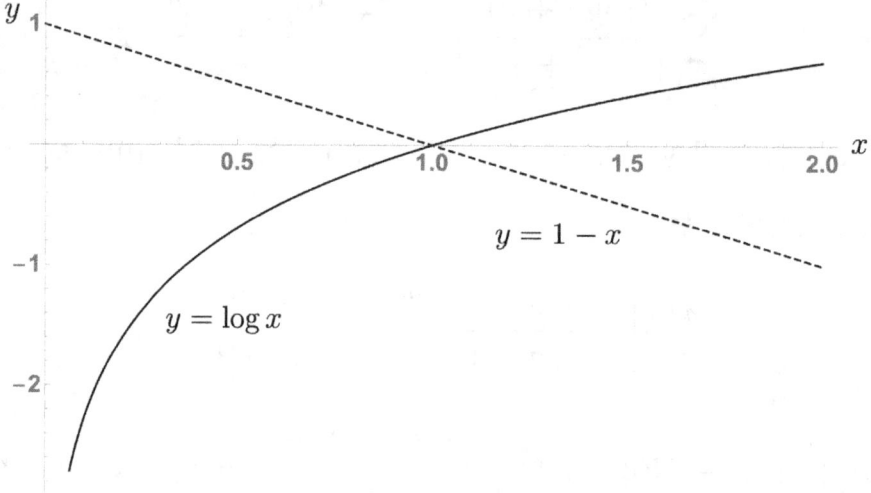

Figure 4.3: Plot of the functions $y = \log x$ (continuous line) and $y = 1 - x$ (dashed line).

$x = 1$, the inequality has the solution

$$0 < x < 1.$$

The second inequality of the system can be written as $\log x >$

4.4 Exercise 4

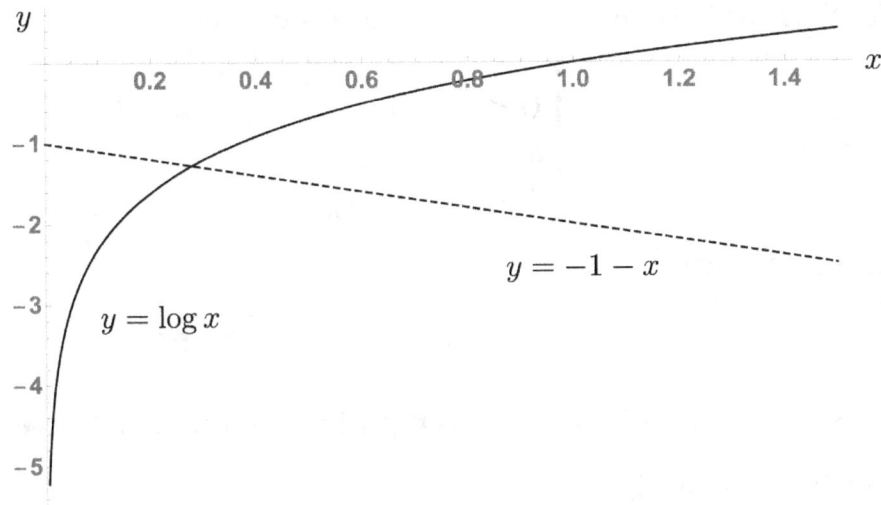

Figure 4.4: Plot of the functions $y = \log x$ (continuous line) and $y = -1 - x$ (dashed line).

$-1 - x$ and we solve it graphically as for the previous one. With reference to Figure 4.4 we call x_0 ($0 < x_0 < 1$) the abscissa of the intersection point between the plots of the functions $y = \log x$ and $y = -1 - x$. The solution of the inequality is

$$x > x_0.$$

We observe that $x_0 < x_1$. The system becomes

$$\begin{cases} 0 < x < 1 \\ x > x_0 \end{cases},$$

whose solution is

$$x_0 < x < 1.$$

The series is **absolutely convergent** and then also **convergent** if $x \in E$ with

$$E \equiv \{x \in \mathbb{R} : x_0 < x < 1\},$$

absolutely divergent if $x \in F$ with

$$F \equiv \{x \in \mathbb{R} : 0 < x < x_0 \lor x > 1\},$$

with unknown behavior if

$$x \in S \equiv \{x_0, 1\}.$$

The values of x for which the series is both absolutely divergent and with positive terms can be obtained intersecting the respective sets A and F. We obtain

$$D \equiv A \cap F = \{x \in \mathbb{R} : x > 1\}.$$

4.4 Exercise 4

For $x \in D$ the series is **divergent**, because the study of absolute divergence coincides with that of simple divergence, being a positive terms series.

If the series is absolutely divergent ($x \in F$) and with alternate sign terms ($x \in B$), i.e.

$$x \in G \equiv B \cap F = \{0 < x < x_0\},$$

we can write

$$\sum_{n=1}^{\infty} \frac{(\log x + x)^n}{n^2 + 3} = \sum_{n=1}^{\infty} (-1)^n \frac{|\log x + x|^n}{n^2 + 3},$$

having used, in this case,

$$\log x + x = -|\log x + x|.$$

We can apply the Leibniz criterion, being an alternating series. Knowing that for $x \in G$ (thanks to the absolute divergence)

$$\lim_{n \to \infty} \left| \frac{a_{n+1}}{a_n} \right| > 1,$$

as previously calculated, the positive terms sequence a_n is not decreasing from a certain n onwards and so the series is **indeterminate** for $x \in G$.

We study the behavior of the initial series for the values $x \in S$ for which the asymptotic ratio criterion gave no informations (i.e. when $|\log x + x| = 1$). We start with $x = x_0$, in this case we have $\log x + x = -1$, and the series becomes

$$\sum_{n=1}^{\infty} \frac{(-1)^n}{n^2 + 3}.$$

We apply the Leibniz criterion by calculating

$$\lim_{n \to \infty} \frac{1}{n^2 + 3} = 0,$$

moreover the positive terms sequence

$$\frac{1}{n^2 + 3}$$

is not increasing, since $n(n+1) = n^2 + 3$, in the denominator, is an increasing sequence, so for $x = x_0$ the series is **convergent**. Similarly, for $x = 1$, where $\log x + x = 1$, we have

$$\sum_{n=1}^{\infty} \frac{1}{n^2 + 3}.$$

4.4 Exercise 4

We apply the ratio criterion with the generalized harmonic series

$$\sum_{n=1}^{\infty} \frac{1}{n^2}$$

which is convergent. Being

$$\frac{1}{n^2+3} < \frac{1}{n^2},$$

we conclude that for $x = 1$, the series is **convergent**.

Summarizing:

The **SERIES** is:

- **CONVERGENT** if

$$x_0 \leq x \leq 1;$$

- **DIVERGENT** if

$$x > 1;$$

- **INDETERMINATE** if

$$0 < x < x_0.$$

Summary scheme:

	0		x_0		1	
♯	♯	IND	C	C	C	D

We can find an approximation of x_0 which represents the solution of the equation $\log x + x + 1 = 0$. We already know graphically that $0 < x_0 < 1$, also x_0 can be seen as a zero of the function $f(x) = \log x + x + 1$. This function is increasing for each $x \in \mathcal{D}$, where $\mathcal{D} = (0, +\infty)$ represents its domain. We calculate the numerical value $f(1/2) = 3/2 - \log 2 \simeq 0.807 > 0$, being positive as well as $f(1) = 2 > 0$ we can conclude that $0 < x_0 < 1/2$. Proceeding by bisection we calculate $f(1/4) = 5/4 - \log 4 \simeq -0.136 < 0$. We deduce that $1/4 < x_0 < 1/2$, we calculate $f(3/8) \simeq 0.394 > 0$ (being $3/8$ the average value between $1/4$ and $1/2$), from which $1/4 < x_0 < 3/8$ that we write like this $0.25 < x_0 < 0.375$. Continuing this way

$$(0.25 + 0.375)/2 \simeq 0.313, \quad f(0.313) \simeq 0.151 > 0,$$

$$0.25 < x_0 < 0.313, \quad (0.25 + 0.313)/2 \simeq 0.282,$$

4.4 Exercise 4

$$f(0.282) \simeq 0.016 > 0, \quad 0.25 < x_0 < 0.282,$$

$$(0.25 + 0.282)/2 \simeq 0.266, \quad f(0.266) \simeq -0.058 < 0,$$

$$0.266 < x_0 < 0.282.$$

You can proceed until you reach the desired approximation. A numerical calculation gives the approximate solution $x_0 \simeq 0.278$.

4.5 Exercise 5

Text

Study the behavior of the series of functions

$$\sum_{n=2}^{\infty} \frac{1}{n^2 \log n} \left(\frac{x^2 - 9}{2x^2 + 3x - 2} \right)^{n \log n}$$

Solution

The basis of the power must be non-negative, because the exponent, i.e. $n \log n$, assumes non-integer values. The denominator must be different from zero. To find the domain we have to solve the following system

$$\begin{cases} \frac{x^2 - 9}{2x^2 + 3x - 2} \geq 0 \\ 2x^2 + 3x - 2 \neq 0 \end{cases}$$

We decompose the second degree polynomials. We have $x^2 - 9 = (x+3)(x-3)$, while the roots of the polynomial $2x^2 + 3x - 2$ can be calculated by solving $2x^2 + 3x - 2 = 0$, $\Delta = 9 + 16 = 25$, $x_{1,2} = (-3 \pm 5)/4$, hence the roots $x_1 =$

4.5 Exercise 5

-2 and $x_2 = 1/2$. We write

$$2x^2 + 3x - 2 = \left(x - \frac{1}{2}\right)(x+2) = (2x-1)(x+2),$$

The system becomes the single inequality

$$\frac{(x+3)(x-3)}{(2x-1)(x+2)} \geq 0,$$

in fact the condition on the denominator is included in the solution of the latter. We study the sign of the four factors separately, we have $x + 3 > 0$ if $x > -3$, $x - 3 > 0$ if $x > 3$, $2x - 1 > 0$ if $x > 1/2$ and $x + 2 > 0$ if $x > -2$. Combining all the results we obtain, as a solution of the inequality,

$$x \leq -3 \ \lor \ -2 < x < \frac{1}{2} \ \lor \ x \geq 3.$$

The domain is

$$\mathcal{D} = \{x \in \mathbb{R} : x \leq -3 \ \lor \ -2 < x < \frac{1}{2} \ \lor \ x \geq 3\}.$$

For $x = \pm 3$ the series has null terms so it is **convergent**. The series has positive terms, we then apply the asymptotic ratio criterion

$$\left|\frac{a_{n+1}}{a_n}\right| = \frac{n^2 \log n}{(n+1)^2 \log(n+1)} \cdot \left(\frac{x^2-9}{2x^2+3x-2}\right)^{(n+1)\log(n+1)}$$
$$\cdot \left(\frac{2x^2+3x-2}{x^2-9}\right)^{n\log n},$$

$$\lim_{n\to\infty}\left|\frac{a_{n+1}}{a_n}\right| = \lim_{n\to\infty} \frac{n^2 \log n}{(n+1)^2 \log(n+1)}$$
$$\cdot \left(\frac{x^2-9}{2x^2+3x-2}\right)^{n\log[(n+1)/n]+\log(n+1)} \cdot$$

We consider the limits

$$\lim_{n\to\infty} \frac{n^2 \log n}{(n+1)^2 \log(n+1)} = 1,$$

$$\lim_{n\to\infty} n\log\left(\frac{n+1}{n}\right) = \lim_{n\to\infty} \log\left[\left(\frac{n+1}{n}\right)^n\right]$$
$$= \lim_{n\to\infty} \log\left[\left(1+\frac{1}{n}\right)^n\right] = \log e = 1.$$

4.5 Exercise 5

The previous limit becomes

$$\lim_{n \to \infty} \left| \frac{a_{n+1}}{a_n} \right| = \lim_{n \to \infty} \left(\frac{x^2 - 9}{2x^2 + 3x - 2} \right)^{\log(n+1)+1}$$

$$= \begin{cases} 0 & \text{if } \frac{x^2-9}{2x^2+3x-2} < 1 \\ \infty & \text{if } \frac{x^2-9}{2x^2+3x-2} > 1 \\ 1 & \text{if } \frac{x^2-9}{2x^2+3x-2} = 1 \end{cases}$$

We solve the inequality

$$\frac{x^2 - 9}{2x^2 + 3x - 2} < 1, \quad \frac{x^2 - 9 - 2x^2 - 3x + 2}{2x^2 + 3x - 2} < 0,$$

$$\frac{x^2 + 3x + 7}{(2x - 1)(x + 2)} > 0.$$

The numerator is always positive (it has complex roots, $\Delta < 0$, and its plot is represented by a parabola with upward concavity). We study the sign of the denominator factors, we have $2x - 1 > 0$ if $x > 1/2$ and $x + 2 > 0$ if $x > -2$. The solution of the inequality is $x < -2 \lor x > \frac{1}{2}$, combining with the domain $\mathcal{D}$ we obtain

$$x < -3 \lor x > 3.$$

In particular we observe that the equation

$$\frac{x^2-9}{2x^2+3x-2}=1, \quad \frac{x^2+3x+7}{(2x-1)(x+2)}=0,$$

has no real solutions.

The series is **convergent** if $x \in E$ with

$$E \equiv \{x \in \mathbb{R} : x < -3 \vee x > 3\},$$

and **divergent** if $x \in F$ with

$$F \equiv \left\{x \in \mathbb{R} : -2 < x < \frac{1}{2}\right\}.$$

Summarizing:

The **SERIES** is:

- **CONVERGENT** if

$$x \leq -3 \vee x \geq 3;$$

- **DIVERGENT** if

$$-2 < x < \frac{1}{2}.$$

4.5 Exercise 5

Summary scheme:

	-3		-2		$1/2$		3	
C	C	$\not\exists$	$\not\exists$	D	$\not\exists$	$\not\exists$	C	C

4.6 Exercise 6

Text

Study the behavior of the series of functions

$$\sum_{n=2}^{\infty} \frac{x+1+nx}{x^2+n^3}$$

Solution

The argument of the series, a sequence of functions, is expressed as a ratio of polynomials in the variable x, where the denominator never vanishes. The domain is therefore $\mathcal{D} = \mathbb{R}$. We consider the absolute series

$$\sum_{n=2}^{\infty} \left| \frac{x+1+nx}{x^2+n^3} \right|,$$

and we apply the asymptotic ratio criterion with the generalized harmonic series

$$\sum_{n=2}^{\infty} \frac{1}{n^\alpha}.$$

4.6 Exercise 6

We have, assuming that $x \neq 0$, case that will be discussed later,

$$\lim_{n \to \infty} \left| \frac{x+1+nx}{x^2+n^3} \right| \cdot n^\alpha = \lim_{n \to \infty} \left| \frac{nx}{n^3} \right| \cdot n^\alpha = |x| \lim_{n \to \infty} n^{\alpha-2}$$

$$= \begin{cases} \infty & \text{if } \alpha > 2 \\ 0 & \text{if } \alpha < 2 \end{cases}$$

For example, we choose, $\alpha = 3/2$. In this case the generalized harmonic series converges and we have

$$\lim_{n \to \infty} \left| \frac{x+1+nx}{x^2+n^3} \right| \cdot n^{3/2} = 0.$$

condition that is true also for $x = 0$. We can conclude that the series is **absolutely convergent**, and then **convergent** for each $x \in \mathcal{D}$. The **SERIES** is:

- **CONVERGENT** if

$$\forall x \in \mathbb{R}.$$

Summary scheme:

	0	
C	C	C

4.7 Exercise 7

Text

Study the behavior of the series of functions

$$\sum_{n=1}^{\infty} \frac{(1-x^2)^{n^2} e^n}{n^2}$$

Solution

There are no restrictions on the values that the x variable can assume, so the domain is $\mathcal{D} = \mathbb{R}$.

Having n^2 the same parity as n, we solve the inequality

$$1 - x^2 > 0.$$

The associated equation $1 - x^2 = 0$ has the solutions $x = \pm 1$, hence

$$-1 < x < 1.$$

So the series is with **positive terms** if $x \in A$ with

$$A \equiv \{x \in \mathbb{R} : -1 < x < 1\},$$

with **alternate sign terms** if $x \in B$ with

$$B \equiv \{x \in \mathbb{R} : x < -1 \lor x > 1\},$$

4.7 Exercise 7

with null terms, and therefore **convergent**, if

$$x \in Z \equiv \{\pm 1\}.$$

We consider the absolute series

$$\sum_{n=1}^{\infty} \left| \frac{(1-x^2)^{n^2} e^n}{n^2} \right| = \sum_{n=1}^{\infty} \frac{|1-x^2|^{n^2} e^n}{n^2}$$

in fact $\forall n > 1$ we have $e^n > 0$ and $n^2 > 0$, and we apply the asymptotic ratio criterion

$$\left| \frac{a_{n+1}}{a_n} \right| = \frac{|1-x^2|^{(n+1)^2} e^{n+1}}{(n+1)^2} \cdot \frac{n^2}{|1-x^2|^{n^2} e^n}$$

$$= |1-x^2|^{2n+1} e \cdot \frac{n^2}{n^2+2n+1},$$

$$\lim_{n \to \infty} \left| \frac{a_{n+1}}{a_n} \right| = e \cdot \lim_{n \to \infty} |1-x^2|^{2n+1} \cdot \frac{n^2}{n^2+2n+1}$$

$$= e \cdot \lim_{n \to \infty} |1-x^2|^{2n+1},$$

i.e.

$$\lim_{n \to \infty} \left| \frac{a_{n+1}}{a_n} \right| = \begin{cases} 0 & \text{if } |1-x^2| < 1 \\ \infty & \text{if } |1-x^2| > 1 \\ e & \text{if } |1-x^2| = 1 \end{cases}.$$

We solve the inequality

$$|1-x^2| < 1,$$

which is equivalent to the system

$$\begin{cases} 1-x^2 < 1 \\ 1-x^2 > -1 \end{cases}$$

The first inequality is $1-x^2 < 1$, $x^2 > 0$ and has the solution

$$x \neq 0,$$

while the second $1-x^2 > -1$, $x^2 - 2 < 0$ has the solution

$$-\sqrt{2} < x < \sqrt{2}.$$

The system becomes

$$\begin{cases} x \neq 0 \\ -\sqrt{2} < x < \sqrt{2} \end{cases},$$

whose solution is

$$-\sqrt{2} < x < 0 \ \lor \ 0 < x < \sqrt{2}.$$

4.7 Exercise 7

The series is **absolutely convergent** and so **convergent** if $x \in E$ with

$$E \equiv \{x \in \mathbb{R} : -\sqrt{2} < x < 0 \vee 0 < x < \sqrt{2}\},$$

absolutely divergent if $x \in F$ with

$$F \equiv \{x \in \mathbb{R} : x \leq -\sqrt{2} \vee x = 0 \vee x \geq \sqrt{2}\}.$$

We search the values of x for which the series is both absolutely divergent and with positive terms, intersecting the respective sets A and F. We obtain

$$D \equiv A \cap F = \{x \in \mathbb{R} : x = 0\}.$$

For $x \in D$, i.e. $x = 0$, the series is **divergent**, because the study of absolute divergence coincides with that of simple divergence, being a positive terms series.

When the series is absolutely divergent ($x \in F$) and with alternate sign terms ($x \in B$), i.e. for

$$x \in G \equiv B \cap F = \{x \leq -\sqrt{2} \vee x \geq \sqrt{2}\},$$

we can write

$$\sum_{n=1}^{\infty} \frac{(1-x^2)^{n^2} e^n}{n^2} = \sum_{n=1}^{\infty} (-1)^n \frac{|1-x^2|^{n^2} e^n}{n^2},$$

having used, in this case,

$$1 - x^2 = -|1 - x^2|,$$

together with the identity $(-1)^{n^2} = (-1)^n$. We can apply the Leibniz criterion, being an alternating series. Knowing that for $x \in G$ (thanks to the absolute divergence)

$$\lim_{n \to \infty} \left| \frac{a_{n+1}}{a_n} \right| > 1,$$

as previously calculated, the positive terms sequence a_n is not decreasing from a certain n onwards and the series is **indeterminate** for $x \in G$.

Summarizing:

The **SERIES** is:

- **CONVERGENT** if

$$-\sqrt{2} < x < 0 \ \vee \ 0 < x < \sqrt{2};$$

4.7 Exercise 7

- **DIVERGENT** if
$$x = 0;$$

- **INDETERMINATE** if
$$x \leq -\sqrt{2} \ \lor \ x \geq \sqrt{2}.$$

Summary scheme:

	$-\sqrt{2}$		0		$\sqrt{2}$	
IND	IND	C	D	C	IND	IND

4.8 Exercise 8

Text

Study the behavior of the series of functions

$$\sum_{n=2}^{\infty} \frac{n\sqrt{n}+1}{n^3 \log n} \left(\frac{x^2+x-12}{x^2-1} \right)^n$$

Solution

To find the domain we solve

$$x^2 - 1 \neq 0, \quad x \neq \pm 1.$$

The domain is $\mathcal{D} = \mathbb{R} - \{\pm 1\}$.

We decompose the polynomials in the variable x. We solve the equation $x^2 + x - 12 = 0$, $\Delta = 1 + 48 = 49$, $x_{1,2} = (-1 \pm 7)/2$, hence the solutions $x_1 = -4$, $x_2 = 3$, so we can write

$$x^2 + x - 12 = (x+4)(x-3),$$

also

$$x^2 - 1 = (x-1)(x+1),$$

4.8 Exercise 8

The series can be written as

$$\sum_{n=2}^{\infty} \frac{n\sqrt{n}+1}{n^3 \log n} \cdot \left(\frac{(x+4)(x-3)}{(x-1)(x+1)} \right)^n.$$

We solve the inequality

$$\frac{(x+4)(x-3)}{(x-1)(x+1)} > 0.$$

We study the sign of the four factors: $x+4 > 0$ if $x > -4$, $x-3 > 0$ if $x > 3$, $x-1 > 0$ if $x > 1$, $x+1 > 0$ if $x > -1$. Combining these results we obtain

$$x < -4 \ \lor \ -1 < x < 1 \ \lor \ x > 3.$$

So the series is with **positive terms** if $x \in A$ with

$$A \equiv \{x \in \mathbb{R} : x < -4 \lor -1 < x < 1 \lor x > 3\},$$

with **alternate sign terms** if $x \in B$ with

$$B \equiv \{x \in \mathbb{R} : -4 < x < -1 \lor 1 < x < 3\},$$

with null terms, and therefore **convergent**, if

$$x \in Z \equiv \{-4, 3\}.$$

We consider the absolute series

$$\sum_{n=2}^{\infty}\left|\frac{n\sqrt{n}+1}{n^3\log n}\cdot\left(\frac{(x+4)(x-3)}{(x-1)(x+1)}\right)^n\right| = \sum_{n=2}^{\infty}\frac{n\sqrt{n}+1}{n^3\log n}\cdot\left|\frac{(x+4)(x-3)}{(x-1)(x+1)}\right|^n,$$

in fact $\forall n > 1$ we have $n\sqrt{n}+1 > 0$ and $n^3\log n > 0$, and we apply the asymptotic ratio criterion

$$\left|\frac{a_{n+1}}{a_n}\right| = \frac{(n+1)\sqrt{n+1}+1}{(n+1)^3\log(n+1)}\cdot\frac{n^3\log n}{n\sqrt{n}+1}\cdot\left|\frac{(x+4)(x-3)}{(x-1)(x+1)}\right|,$$

$$\lim_{n\to\infty}\left|\frac{a_{n+1}}{a_n}\right| = \left|\frac{(x+4)(x-3)}{(x-1)(x+1)}\right|\cdot\lim_{n\to\infty}\frac{(n+1)\sqrt{n+1}+1}{n\sqrt{n}+1}$$

$$\cdot\frac{n^3\log n}{(n+1)^3\log(n+1)} = \left|\frac{(x+4)(x-3)}{(x-1)(x+1)}\right|.$$

We solve the inequality

$$\left|\frac{(x+4)(x-3)}{(x-1)(x+1)}\right| < 1,$$

which is equivalent to the system

$$\begin{cases}\frac{(x+4)(x-3)}{(x-1)(x+1)} < 1 \\ \frac{(x+4)(x-3)}{(x-1)(x+1)} > -1\end{cases}, \quad \begin{cases}\frac{x^2+x-12-x^2+1}{(x-1)(x+1)} < 0 \\ \frac{x^2+x-12+x^2-1}{(x-1)(x+1)} > 0\end{cases},$$

4.8 Exercise 8

$$\begin{cases} \frac{x-11}{(x-1)(x+1)} < 0 \\ \frac{2x^2+x-13}{(x-1)(x+1)} > 0 \end{cases}.$$

The first inequality is

$$\frac{x-11}{(x-1)(x+1)} < 0$$

We study the sign of the three factors: $x-11 > 0$ if $x > 11$, $x-1 > 0$ if $x > 1$, $x+1 > 0$ if $x > -1$ and, combining them, we get the solution

$$x < -1 \;\;\lor\;\; 1 < x < 11,$$

while the second inequality is

$$\frac{2x^2+x-13}{(x-1)(x+1)} > 0.$$

We solve the equation $2x^2+x-13=0$, $\Delta=1+104=105$, from which the roots

$$x_{1,2} = \frac{-1 \pm \sqrt{105}}{4}.$$

Studying the signs of the three factors we obtain: $2x^2 + x - 13 > 0$ if

$$x < \frac{-1-\sqrt{105}}{4} \quad \vee \quad x > \frac{-1+\sqrt{105}}{4},$$

$x - 1 > 0$ if $x > 1$, $x + 1 > 0$ if $x > -1$ and, by combining them, we have the solution

$$x < \frac{-1-\sqrt{105}}{4} \quad \vee \quad -1 < x < 1 \quad \vee \quad x > \frac{-1+\sqrt{105}}{4}.$$

The system becomes

$$\begin{cases} x < -1 \ \vee \ 1 < x < 11 \\ x < \frac{-1-\sqrt{105}}{4} \ \vee \ -1 < x < 1 \ \vee \ x > \frac{-1+\sqrt{105}}{4} \end{cases},$$

whose solution is

$$x < \frac{-1-\sqrt{105}}{4} \quad \vee \quad \frac{-1+\sqrt{105}}{4} < x < 11.$$

The series is **absolutely convergent** and then it is also **convergent** if $x \in E$ with

$$E \equiv \left\{ x \in \mathbb{R} : x < \frac{-1-\sqrt{105}}{4} \vee \frac{-1+\sqrt{105}}{4} < x < 11 \right\},$$

4.8 Exercise 8

absolutely divergent if $x \in F$ with

$$F \equiv \left\{ x \in \mathbb{R} : \frac{-1-\sqrt{105}}{4} < x < -1 \right.$$
$$\left. \vee \ -1 < x < 1 \vee x > \frac{-1+\sqrt{105}}{4} \vee x > 11 \right\},$$

with unknown behavior if

$$x \in S \equiv \left\{ \frac{-1-\sqrt{105}}{4}, \frac{-1+\sqrt{105}}{4}, 11 \right\}.$$

We search the values of x for which the series is both absolutely divergent and with positive terms, intersecting the respective sets A ed F. We obtain

$$D \equiv A \cap F = \{ x \in \mathbb{R} : -1 < x < 1 \vee x > 11 \}.$$

For $x \in D$ the series is **divergent**, because the study of absolute divergence coincides with that of simple divergence, being a positive terms series.

If the series is absolutely divergent ($x \in F$) and with alternate sign terms ($x \in B$), i.e. for

$$x \in G \equiv B \cap F = \left\{ \frac{-1-\sqrt{105}}{4} < x < -1 \right.$$
$$\left. \vee \quad 1 < x < \frac{-1+\sqrt{105}}{4} \right\},$$

we can write

$$\sum_{n=2}^{\infty} \frac{n\sqrt{n}+1}{n^3 \log n} \cdot \left(\frac{(x+4)(x-3)}{(x-1)(x+1)} \right)^n = \sum_{n=2}^{\infty} (-1)^n \frac{n\sqrt{n}+1}{n^3 \log n}$$
$$\cdot \left| \frac{(x+4)(x-3)}{(x-1)(x+1)} \right|^n,$$

having used, in this case,

$$\frac{(x+4)(x-3)}{(x-1)(x+1)} = - \left| \frac{(x+4)(x-3)}{(x-1)(x+1)} \right|.$$

We can apply the Leibniz criterion, being an alternating series. Knowing that for $x \in G$ (thanks to the absolute divergence)

$$\lim_{n \to \infty} \left| \frac{a_{n+1}}{a_n} \right| > 1,$$

as previously calculated, the positive terms sequence a_n is not decreasing from a certain n onwards and therefore the

4.8 Exercise 8

series is **indeterminate** for $x \in G$.

We determine the behavior of the initial series for the values $x \in S$ for which the asymptotic ratio criterion has not provided informations, i.e. when

$$\left|\frac{(x+4)(x-3)}{(x-1)(x+1)}\right| = 1.$$

We start with

$$x = \frac{-1 \pm \sqrt{105}}{4},$$

in this case we have

$$\frac{(x+4)(x-3)}{(x-1)(x+1)} = -1$$

and the series becomes

$$\sum_{n=2}^{\infty} (-1)^n \frac{n\sqrt{n}+1}{n^3 \log n}.$$

We apply the Leibniz criterion, firstly calculating

$$\lim_{n \to \infty} \frac{n\sqrt{n}+1}{n^3 \log n} = \lim_{n \to \infty} \frac{\sqrt{n}}{n^2 \log n} = \lim_{n \to \infty} \frac{1}{n^{3/2} \log n} = 0.$$

We study the monotony of the positive terms series

$$b_n = \frac{n\sqrt{n}+1}{n^3 \log n},$$

we solve, remembering that $n \geq 2$, the inequality

$$b_{n+1} - b_n = \frac{(n+1)\sqrt{n+1}+1}{(n+1)^3 \log(n+1)} - \frac{n\sqrt{n}+1}{n^3 \log n} < 0,$$

$$\left((n+1)\sqrt{n+1}+1\right)(n^3 \log n) - \left((n+1)^3 \log n + 1\right)$$
$$\cdot (n\sqrt{n}+1) < 0$$

$$\left((n+1)^3 \log n + 1\right)(n\sqrt{n}+1) > \left((n+1)\sqrt{n+1}+1\right)$$
$$\cdot (n^3 \log n)$$

$$\log\left(e^{n\sqrt{n}+1}\right) + \log\left(n^{(n+1)^3(n\sqrt{n}+1)}\right)$$
$$> \log\left(n^{n^3[(n+1)\sqrt{n+1}+1]}\right)$$

$$\log\left(n^{(n+1)^3(n\sqrt{n}+1)} e^{n\sqrt{n}+1}\right) > \log\left(n^{n^3[(n+1)\sqrt{n+1}+1]}\right)$$

$$n^{(n+1)^3(n\sqrt{n}+1)} e^{n\sqrt{n}+1} > n^{n^3[(n+1)\sqrt{n+1}+1]}$$

$$n^{(n^4+3n^3+3n^2+n)\sqrt{n}+3n^2+3n+1} e^{n\sqrt{n}+1} > n^{(n^4+n^3)\sqrt{n+1}}$$

We show that

$$n^{(n^4+3n^3+3n^2+n)\sqrt{n}} > n^{(n^4+n^3)\sqrt{n+1}}$$

4.8 Exercise 8

from which

$$(n^4 + 3n^3 + 3n^2 + n)\sqrt{n} > (n^4 + n^3)\sqrt{n+1},$$

$$n^3(n^3 + 3n^2 + 3n + 1)^2 > (n+1)^2(n+1)n^6$$

$$(n+1)^3(n+1)^3 > (n+1)^3 n^3, \quad (n+1)^3 > n^3,$$

true $\forall n$. We have demonstrated that $b_{n+1} - b_n < 0$ and that the sequence with positive terms b_n is decreasing (and in particular not increasing) $\forall n \geq 2$.

We can conclude that for

$$x = \frac{-1 \pm \sqrt{105}}{4},$$

the series is **convergent**. In Figure 4.5 we show the plot of the sequence b_n.

Similarly, for $x = 11$, where

$$\frac{(x+4)(x-3)}{(x-1)(x+1)} = 1,$$

we have

$$\sum_{n=2}^{\infty} \frac{n\sqrt{n}+1}{n^3 \log n}.$$

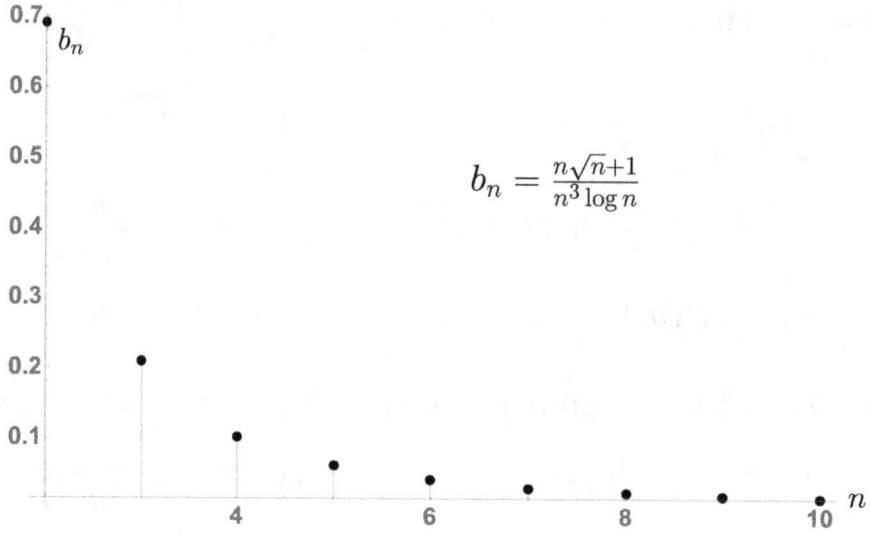

Figure 4.5: Plot of the sequence b_n for $n \in [2, 10] \subset \mathbb{R}$.

We apply the asymptotic ratio criterion with the generalized harmonic series

$$\sum_{n=1}^{\infty} \frac{1}{n^\alpha},$$

4.8 Exercise 8

we calculate

$$\lim_{n\to\infty} \frac{n\sqrt{n}+1}{n^3 \log n} \frac{1}{1/n^\alpha} = \lim_{n\to\infty} \frac{(n\sqrt{n}+1)n^\alpha}{n^3 \log n} = \lim_{n\to\infty} \frac{n^{\alpha-3/2}}{\log n}$$

$$= \begin{cases} \infty & \text{if } \alpha > 3/2 \\ 0 & \text{if } \alpha \leq 3/2 \end{cases}.$$

We observe that it is sufficient to choose $\alpha = 1/2$ to conclude that, since the generalized harmonic series is divergent

$$\sum_{n=1}^{\infty} \frac{1}{n^{1/2}},$$

then, for $x = 11$, the initial series is **divergent**.

Summarizing:

The **SERIES** is:

- **CONVERGENT** if

$$x \leq \frac{-1-\sqrt{105}}{4} \quad \vee \quad \frac{-1+\sqrt{105}}{4} \leq x < 11;$$

- **DIVERGENT** if

$$-1 < x < 1 \quad \vee \quad x \geq 11;$$

- **INDETERMINATE** if

$$\frac{-1-\sqrt{105}}{4} < x < -1 \;\; \lor \;\; 1 < x < \frac{-1+\sqrt{105}}{4}.$$

Summary scheme:

	$\frac{-1-\sqrt{105}}{4}$		-1		1		$\frac{-1+\sqrt{105}}{4}$		11	
C		C	IND	∄	D	∄	IND	C	C	D	D

4.9 Exercise 9

Text

Study the behavior of the series of functions

$$\sum_{n=1}^{\infty} \frac{(2-x^2)^n}{1+n+n^2}$$

Solution

There are no restrictions on the values that the x variable can take, so the domain is $\mathcal{D} = \mathbb{R}$.

We solve the inequality

$$2 - x^2 > 0.$$

The associated equation $2 - x^2 = 0$ has the roots $x = \pm\sqrt{2}$, hence

$$-\sqrt{2} < x < \sqrt{2}.$$

So the series is with **positive terms** if $x \in A$ with

$$A \equiv \{x \in \mathbb{R} : -\sqrt{2} < x < \sqrt{2}\},$$

with **alternate sign terms** if $x \in B$ with

$$B \equiv \{x \in \mathbb{R} : x < -\sqrt{2} \vee x > \sqrt{2}\},$$

with null terms, and therefore **convergent**, if

$$x \in Z \equiv \{\pm\sqrt{2}\}.$$

We consider the absolute series

$$\sum_{n=1}^{\infty} \left|\frac{(2-x^2)^n}{1+n+n^2}\right| = \sum_{n=1}^{\infty} \frac{|2-x^2|^n}{1+n+n^2}$$

in fact $\forall n > 1$ we have $1+n+n^2 > 0$, and we apply the asymptotic ratio criterion

$$\left|\frac{a_{n+1}}{a_n}\right| = \frac{|2-x^2|^{n+1}}{1+n+1+(n+1)^2} \cdot \frac{1+n+n^2}{|2-x^2|^n}$$

$$= |2-x^2| \cdot \frac{1+n+n^2}{3+3n+n^2},$$

$$\lim_{n\to\infty} \left|\frac{a_{n+1}}{a_n}\right| = |2-x^2| \cdot \lim_{n\to\infty} \left(\frac{1+n+n^2}{3+3n+n^2}\right) = |2-x^2|.$$

We solve the inequality

$$|2-x^2| < 1,$$

which is equivalent to the system

$$\begin{cases} 2-x^2 < 1 \\ 2-x^2 > -1 \end{cases}$$

4.9 Exercise 9

The first inequality is $2 - x^2 < 1$, $1 - x^2 < 0$ and has the solution

$$x < -1 \quad \lor \quad x > 1,$$

while the second $2 - x^2 > -1$, $3 - x^2 > 0$ has the solution

$$-\sqrt{3} < x < \sqrt{3}.$$

The system becomes

$$\begin{cases} x < -1 \quad \lor \quad x > 1 \\ -\sqrt{3} < x < \sqrt{3} \end{cases},$$

whose solution is

$$-\sqrt{3} < x < -1 \quad \lor \quad 1 < x < \sqrt{3}.$$

The series is **absolutely convergent** and then it is also **convergent** if $x \in E$ with

$$E \equiv \{x \in \mathbb{R} : -\sqrt{3} < x < -1 \lor 1 < x < \sqrt{3}\},$$

absolutely divergent if $x \in F$ with

$$F \equiv \{x \in \mathbb{R} : x < -\sqrt{3} \lor -1 < x < 1 \lor x > \sqrt{3}\},$$

with unknown behavior if

$$x \in S \equiv \{\pm 1, \pm\sqrt{3}\}.$$

We search the values of x for which the series is both absolutely divergent and with positive terms that can be obtained intersecting the respective sets A and F. We have

$$D \equiv A \cap F = \{x \in \mathbb{R} : -1 < x < 1\}.$$

For $x \in D$ the series is **divergent**, because the study of absolute divergence coincides with that of simple divergence, being a positive terms series.

In the case where the series is absolutely divergent ($x \in F$) and with alternate sign terms ($x \in B$), i.e. for

$$x \in G \equiv B \cap F = \{x < -\sqrt{3} \vee x > \sqrt{3}\},$$

we can write

$$\sum_{n=1}^{\infty} \frac{(2-x^2)^n}{1+n+n^2} = \sum_{n=1}^{\infty} (-1)^n \frac{|2-x^2|^n}{1+n+n^2},$$

4.9 Exercise 9

having used, in this case,

$$2 - x^2 = -|2 - x^2|.$$

We can apply the Leibniz criterion, being an alternating series. Knowing that for $x \in G$ (thanks to the absolute divergence)

$$\lim_{n \to \infty} \left| \frac{a_{n+1}}{a_n} \right| > 1,$$

as previously calculated, the positive terms sequence a_n is not decreasing from a certain n onwards and so the series is **indeterminate** for $x \in G$.

We determine the behavior of the initial series for the values $x \in S$ for which the asymptotic ratio criterion did not provide information (i.e. when $|2 - x^2| = 1$). For $x = \pm\sqrt{3}$, we have $2 - x^2 = -1$, and the series becomes

$$\sum_{n=1}^{\infty} \frac{(-1)^n}{1 + n + n^2}.$$

We apply the Leibniz criterion by calculating

$$\lim_{n \to \infty} \frac{1}{1 + n + n^2} = 0,$$

moreover the positive terms sequence

$$\frac{1}{1+n+n^2}$$

is not increasing, being $n(n+1) = 1+n+n^2$, in the denominator, an increasing sequence, and we conclude that for $x = \pm\sqrt{3}$ the series is **convergent**. Similarly, for $x = \pm 1$, where $2 - x^2 = 1$, we have

$$\sum_{n=1}^{\infty} \frac{1}{1+n+n^2}.$$

We apply the ratio criterion with the convergent generalized harmonic series

$$\sum_{n=1}^{\infty} \frac{1}{n^2},$$

in fact, being

$$\frac{1}{1+n+n^2} < \frac{1}{n^2},$$

we can conclude that for $x = \pm 1$ the series is **convergent**.

Summarizing:

The **SERIES** is:

4.9 Exercise 9

- **CONVERGENT** if

$$-\sqrt{3} \leq x \leq -1 \ \lor \ 1 \leq x \leq \sqrt{3};$$

- **DIVERGENT** if

$$-1 < x < 1;$$

- **INDETERMINATE** if

$$x < -\sqrt{3} \ \lor \ x > \sqrt{3}.$$

Summary scheme:

	$-\sqrt{3}$		-1		1		$\sqrt{3}$	
IND	C	C	C	D	C	C	C	IND

4.10 Exercise 10

Text

Study the behavior of the series of functions

$$\sum_{n=2}^{\infty} \frac{(\sqrt{x}-3)^n}{\log n}$$

Solution

The radicand must be a non-negative quantity, i.e. $x \geq 0$, hence the domain is $\mathcal{D} = \{x \in \mathbb{R} : x \geq 0\}$.

We solve the inequality

$$\sqrt{x} - 3 > 0$$

which is equivalent to the system

$$\begin{cases} x > 9 \\ x \geq 0 \end{cases},$$

hence the solution is $x > 9$.

The series is with **positive terms** if $x \in A$ with

$$A \equiv \{x \in \mathbb{R} : x > 9\},$$

4.10 Exercise 10

with **alternate sign terms** if $x \in B$ with

$$B \equiv \{x \in \mathbb{R} : 0 \leq x < 9\},$$

with null terms, and therefore **convergent**, if

$$x \in Z \equiv \{9\}.$$

We consider the absolute series

$$\sum_{n=2}^{\infty} \left| \frac{(\sqrt{x}-3)^n}{\log n} \right| = \sum_{n=2}^{\infty} \frac{|\sqrt{x}-3|^n}{\log n},$$

in fact $\forall n > 1$ we have $\log n > 0$, and we apply the asymptotic ratio criterion

$$\left| \frac{a_{n+1}}{a_n} \right| = \frac{|\sqrt{x}-3|^{n+1}}{\log(n+1)} \cdot \frac{\log n}{|\sqrt{x}-3|^n} = |\sqrt{x}-3| \cdot \frac{\log n}{\log(n+1)},$$

$$\lim_{n \to \infty} \left| \frac{a_{n+1}}{a_n} \right| = |\sqrt{x}-3| \cdot \lim_{n \to \infty} \frac{\log n}{\log(n+1)} = |\sqrt{x}-3|.$$

We solve the inequality

$$|\sqrt{x}-3| < 1,$$

which is equivalent to the system

$$\begin{cases} \sqrt{x}-3 < 1 \\ \sqrt{x}-3 > -1 \end{cases}, \quad \begin{cases} \sqrt{x} < 4 \\ \sqrt{x} > 2 \end{cases}$$

The system can be written as

$$\begin{cases} x < 16 \\ x > 4 \\ x \geq 0 \end{cases},$$

and has the solution

$$4 < x < 16.$$

The series is **absolutely convergent** and then also **convergent** if $x \in E$ with

$$E \equiv \{x \in \mathbb{R} : 4 < x < 16\},$$

absolutely divergent if $x \in F$ with

$$F \equiv \{x \in \mathbb{R} : 0 \leq x < 4 \lor x > 16\},$$

4.10 Exercise 10

with unknown behavior if

$$x \in S \equiv \{4, 16\}.$$

We search the values of x for which the series is both absolutely divergent and with positive terms that can be obtained intersecting the respective sets A and F. We have

$$D \equiv A \cap F = \{x \in \mathbb{R} : x > 16\}.$$

For $x \in D$ the series is **divergent**, because the study of absolute divergence coincides with that of simple divergence, being a positive terms series.

In the case where the series is absolutely divergent ($x \in F$) and with alternate sign terms ($x \in B$), i.e. for

$$x \in G \equiv B \cap F = \{0 \leq x < 4\},$$

we can write

$$\sum_{n=2}^{\infty} \frac{(\sqrt{x}-3)^n}{\log n} = \sum_{n=2}^{\infty} (-1)^n \frac{|\sqrt{x}-3|^n}{\log n},$$

having used, in this case,
$$\sqrt{x}-3 = -|\sqrt{x}-3|.$$

We can apply the Leibniz criterion, being an alternating series. Knowing that for $x \in G$ (thanks to the absolute divergence)
$$\lim_{n \to \infty} \left|\frac{a_{n+1}}{a_n}\right| > 1,$$
as previously calculated, the positive terms sequence a_n is not decreasing from a certain n onwards and so the series is **indeterminate** for $x \in G$.

We determine the behavior of the initial series for the values $x \in S$ for which the asymptotic ratio criterion has not provided informations (ie when $|\sqrt{x}-3| = 1$). We put $x = 4$, in this case we have $\sqrt{x}-3 = -1$, and the series becomes
$$\sum_{n=2}^{\infty} \frac{(-1)^n}{\log n}.$$
We apply the Leibniz criterion by calculating
$$\lim_{n \to \infty} \frac{1}{\log n} = 0,$$

4.10 Exercise 10

moreover the positive terms sequence

$$\frac{1}{\log n}$$

is not increasing, since $n(n+1) = \log n$, in the denominator, is an increasing sequence, and for $x = 4$ the series is **convergent**. Similarly, for $x = 16$, where $\sqrt{x} - 3 = 1$, we have

$$\sum_{n=2}^{\infty} \frac{1}{\log n}.$$

We apply the asymptotic ratio criterion, with the divergent harmonic series

$$\sum_{n=2}^{\infty} \frac{1}{n},$$

calculating

$$\lim_{n \to \infty} \frac{1/\log n}{1/n} = \lim_{n \to \infty} \frac{n}{\log n} = \infty.$$

We conclude that for $x = 16$ the initial series is **divergent**.

Summarizing:

The **SERIES** is:

- **CONVERGENT** if

$$4 \leq x < 16;$$

- **DIVERGENT** if

$$x \geq 16;$$

- **INDETERMINATE** if

$$0 \leq x < 4.$$

Summary scheme:

$\sharp$	0		4		16	
	IND	IND	C	C	D	D

4.11 Exercise 11

Text

Study the behavior of the series of functions

$$\sum_{n=2}^{\infty} \frac{(x-1)^{n^2} \log n}{n^n}$$

Solution

There is no limitation on the values that the x variable can assume, so the domain is $\mathcal{D} = \mathbb{R}$.

The parity of n^2 is the same as that of n, so we solve the inequality $x - 1 > 0$, whose solution is,

$$x > 1$$

The series is with **positive terms** if $x \in A$ with

$$A \equiv \{x \in \mathbb{R} : x > 1\},$$

with **alternate sign terms** if $x \in B$ with

$$B \equiv \{x \in \mathbb{R} : x < 1\},$$

with null terms, and therefore **convergent**, if

$$x \in Z \equiv \{1\}.$$

We consider the absolute series

$$\sum_{n=1}^{\infty} \left| \frac{(x-1)^{n^2} \log n}{n^n} \right| = \sum_{n=1}^{\infty} \frac{|x-1|^{n^2} \log n}{n^n}$$

in fact $\forall n > 1$ we have $\log n > 0$ e $n^n > 0$, and we apply the asymptotic ratio criterion

$$\left| \frac{a_{n+1}}{a_n} \right| = \frac{|x-1|^{(n+1)^2} \log(n+1)}{(n+1)^{n+1}} \cdot \frac{n^n}{|x-1|^{n^2} \log n}$$

$$= \frac{|x-1|^{2n+1} n^n \log(n+1)}{(n+1)(n+1)^n \log n},$$

$$\lim_{n \to \infty} \left| \frac{a_{n+1}}{a_n} \right| = \lim_{n \to \infty} \left(\frac{|x-1|^{2n+1} n^n \log(n+1)}{(n+1)(n+1)^n \log n} \right)$$

$$= \lim_{n \to \infty} \left(\frac{|x-1|^{2n+1} n^n}{(n+1)(n+1)^n} \right),$$

where we used

$$\lim_{n \to \infty} \frac{\log(n+1)}{\log n} = 1.$$

4.11 Exercise 11

We consider the limit

$$\lim_{n\to\infty} \frac{n^n}{(n+1)^n} = \lim_{n\to\infty} \left(\frac{n}{n+1}\right)^n = \lim_{n\to\infty} \left(1+\frac{1}{n}\right)^{-n}$$

$$= \left[\lim_{n\to\infty} \left(1+\frac{1}{n}\right)^n\right]^{-1} = \frac{1}{e}.$$

The previous limit becomes

$$\lim_{n\to\infty} \left|\frac{a_{n+1}}{a_n}\right| = \frac{1}{e} \lim_{n\to\infty} \left(\frac{|x-1|^{2n+1}}{n+1}\right) = \begin{cases} 0 & \text{if } |x-1| \le 1 \\ \infty & \text{if } |x-1| > 1 \end{cases},$$

We solve the inequality

$$|x-1| < 1,$$

which is equivalent to the system

$$\begin{cases} x-1 < 1 \\ x-1 > -1 \end{cases}, \quad \begin{cases} x < 2 \\ x > 0 \end{cases},$$

whose solution is

$$0 < x < 2.$$

The series is **absolutely convergent** and so **convergent** if $x \in E$ with

$$E \equiv \{x \in \mathbb{R} : 0 \leq x \leq 2\},$$

absolutely divergent if $x \in F$ with

$$F \equiv \{x \in \mathbb{R} : x < 0 \lor x > 2\}.$$

We search the values of x for which the series is both absolutely divergent and with positive terms that can be obtained intersecting the respective sets A and F. We obtain

$$D \equiv A \cap F = \{x \in \mathbb{R} : x > 2\}.$$

For $x \in D$ the series is **divergent**, because the study of absolute divergence coincides with that of simple divergence, being a positive terms series.

If the series is absolutely divergent ($x \in F$) and with alternate sign terms ($x \in B$), i.e. for

$$x \in G \equiv B \cap F = \{x < 0\},$$

4.11 Exercise 11

we can write

$$\sum_{n=2}^{\infty} \frac{(x-1)^{n^2} \log n}{n^n} = \sum_{n=2}^{\infty} \frac{(-1)^n |x-1|^{n^2} \log n}{n^n},$$

having used $(-1)^{n^2} = (-1)^n$ and, in this case,

$$x - 1 = -|x-1|.$$

We can apply the Leibniz criterion, being an alternating series. Knowing that for $x \in G$ (thanks to the absolute divergence)

$$\lim_{n \to \infty} \left| \frac{a_{n+1}}{a_n} \right| > 1,$$

as previously calculated, the positive terms sequence a_n is not decreasing from a certain n onwards and the series is **indeterminate** for $x \in G$.

Summarizing:

The **SERIES** is:

- **CONVERGENT** if

$$0 \leq x \leq 2;$$

- **DIVERGENT** if

$$x > 2;$$

- **INDETERMINATE** if

$$x < 0.$$

Summary scheme:

		0		2	
IND	C	C	C	D	

4.12 Exercise 12

Text

Study the behavior of the series of functions

$$\sum_{n=2}^{\infty} \frac{(x^2 + \log x)^n}{n^{3/2} \log n}$$

Solution

For the presence of $\log x$ we have $x > 0$ as a condition of existence. The domain is $\mathcal{D} = \mathbb{R}^+$.

We solve the inequality

$$x^2 + \log x > 0.$$

In this case, since we cannot solve it analytically, we must proceed graphically, writing $x^2 > -\log x$ and drawing the plot of the function $y = x^2$ and that of the function $y = -\log x$, as shown in Figure 4.6.

We observe that the graphs intersect only at the abscissa point x_1 with $0 < x_1 < 1$. The inequality is satisfied for $x > x_1$.

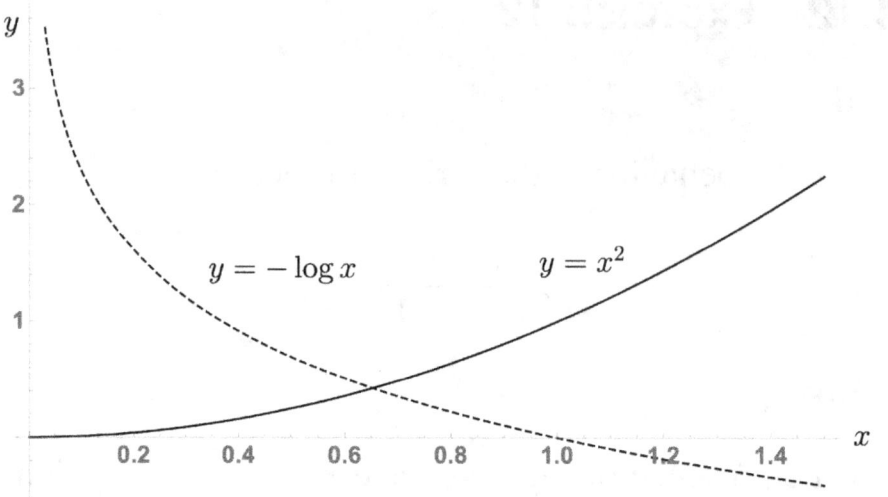

Figure 4.6: Plot of the functions $y = x^2$ (continuous line) and $y = -\log x$ (dashed line).

So the series is with **positive terms** if $x \in A$ with

$$A \equiv \{x \in \mathbb{R} : x > x_1\};$$

with **alternate sign terms** if $x \in B$ with

$$B \equiv \{x \in \mathbb{R} : 0 < x < x_1\};$$

with null terms, and therefore **convergent**, if

$$x \in Z \equiv \{x_1\}.$$

4.12 Exercise 12

We consider the absolute series

$$\sum_{n=2}^{\infty} \left| \frac{(x^2 + \log x)^n}{n^{3/2} \log n} \right| = \sum_{n=2}^{\infty} \frac{|x^2 + \log x|^n}{n^{3/2} \log n}$$

in fact $\forall n > 1$ we have $n^{3/2} \log n > 0$, and we apply the asymptotic ratio criterion

$$\left| \frac{a_{n+1}}{a_n} \right| = \frac{|x^2 + \log x|^{n+1}}{(n+1)^{3/2} \log(n+1)} \cdot \frac{n^{3/2} \log n}{|x^2 + \log x|^n}$$
$$= |x^2 + \log x| \frac{n^{3/2} \log n}{(n+1)^{3/2} \log(n+1)},$$

hence, being

$$\lim_{n \to \infty} \frac{n^{3/2} \log n}{(n+1)^{3/2} \log(n+1)} = 1,$$

$$\lim_{n \to \infty} \left| \frac{a_{n+1}}{a_n} \right| = |x^2 + \log x| \lim_{n \to \infty} \left(\frac{n^{3/2} \log n}{(n+1)^{3/2} \log(n+1)} \right)$$
$$= |x^2 + \log x|.$$

We solve the inequality

$$|x^2 + \log x| < 1,$$

which is equivalent to the system

$$\begin{cases} x^2 + \log x < 1 \\ x^2 + \log x > -1 \end{cases}$$

The first inequality can be written as $x^2 < 1 - \log x$ and can be solved graphically. The plots of the functions $y = x^2$ and $y = 1 - \log x$ are shown in Figure 4.7.
The solution of the associated equation $x^2 = 1 - \log x$ is $x = 1$, the inequality has the solution

$$0 < x < 1.$$

The second inequality of the system can be written as $x^2 > -1 - \log x$ and we solve it graphically as we did before. With reference to Figure 4.8 we define x_0 ($0 < x_0 < 1$) the abscissa of the intersection point between the graphs of the functions $y = x^2$ and $y = -1 - \log x$. The solution of the inequality is

$$x > x_0.$$

4.12 Exercise 12

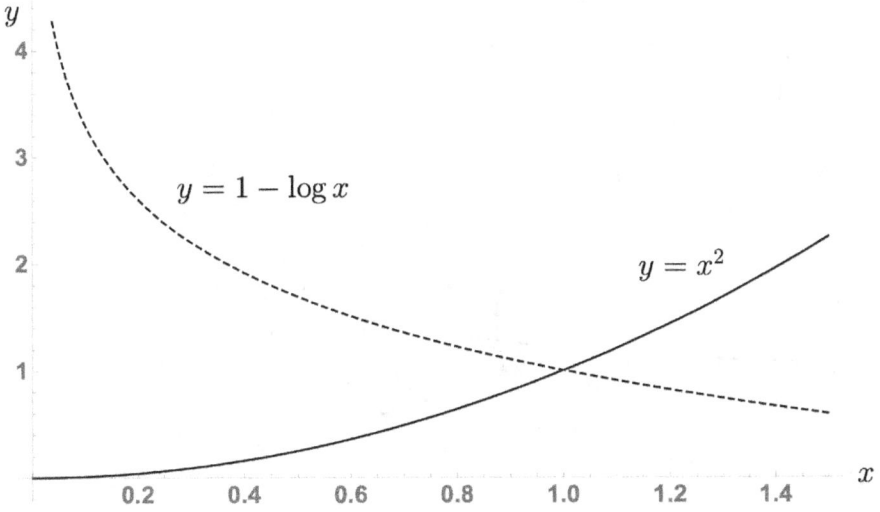

Figure 4.7: Plot of the functions $y = x^2$ (continuous line) and $y = 1 - \log x$ (dashed line).

We also observe that $x_0 < x_1$. The system becomes

$$\begin{cases} 0 < x < 1 \\ x > x_0 \end{cases},$$

whose solution is

$$x_0 < x < 1.$$

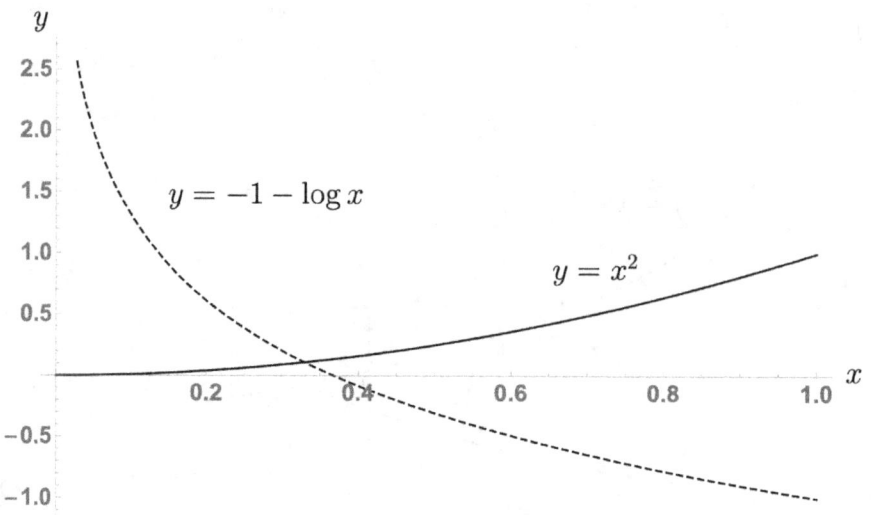

Figure 4.8: Plot of the functions $y = x^2$ (continuous line) and $y = -1 - \log x$ (dashed line).

The series is **absolutely convergent** and then also **convergent** if $x \in E$ with

$$E \equiv \{x \in \mathbb{R} : x_0 < x < 1\},$$

absolutely divergent if $x \in F$ with

$$F \equiv \{x \in \mathbb{R} : 0 < x < x_0 \vee x > 1\},$$

4.12 Exercise 12

with unknown behavior if

$$x \in S \equiv \{x_0, 1\}.$$

We search the values of x for which the series is both absolutely divergent and with positive terms that can be obtained intersecting the respective sets A and F. We obtain

$$D \equiv A \cap F = \{x \in \mathbb{R} : x > 1\}.$$

For $x \in D$ the series is **divergent**, because the study of absolute divergence coincides with that of simple divergence, being a positive terms series.

If the series is absolutely divergent ($x \in F$) and with alternate sign terms ($x \in B$), i.e. for

$$x \in G \equiv B \cap F = \{0 < x < x_0\},$$

we can write

$$\sum_{n=2}^{\infty} \frac{(x^2 + \log x)^n}{n^{3/2} \log n} = \sum_{n=2}^{\infty} (-1)^n \frac{|x^2 + \log x|^n}{n^{3/2} \log n},$$

having used, in this case,
$$x^2 + \log x = -|x^2 + \log x|.$$

We can apply the Leibniz criterion, being an alternating series. Knowing that for $x \in G$ (thanks to the absolute divergence)
$$\lim_{n \to \infty} \left| \frac{a_{n+1}}{a_n} \right| > 1,$$
as previously calculated, the positive terms sequence a_n is not decreasing from a certain n onwards and the series is **indeterminate** for $x \in G$.

We determine the behavior of the initial series for the values $x \in S$ for which the asymptotic ratio criterion has provided no informations (i.e. when $|x^2 + \log x| = 1$). We consider $x = x_0$, in this case we have $x^2 + \log x = -1$, and the series becomes
$$\sum_{n=2}^{\infty} \frac{(-1)^n}{n^{3/2} \log n}.$$

We apply the Leibniz criterion by calculating
$$\lim_{n \to \infty} \frac{1}{n^{3/2} \log n} = 0,$$

4.12 Exercise 12

moreover the positive terms sequence

$$\frac{1}{n^{3/2} \log n}$$

is not increasing, since $n^{3/2} \log n$, in the denominator, is an increasing sequence, and for $x = x_0$ the series is **convergent**. Similarly, for $x = 1$, where $x^2 + \log x = 1$, we have

$$\sum_{n=2}^{\infty} \frac{1}{n^{3/2} \log n}.$$

We apply the ratio criterion with the convergent generalized harmonic series

$$\sum_{n=1}^{\infty} \frac{1}{n^{3/2}}$$

Being, $\forall n \geq 3$,

$$\frac{1}{n^{3/2} \log n} < \frac{1}{n^{3/2}},$$

the term n-th of the series is lower than the corresponding term of the generalized harmonic series (from a certain n onwards, in this case 3), we conclude that for $x = 1$, the series is **convergent**. In fact, remember that changing or deleting a finite number of terms from a series does not change its behavior.

Summarizing:

The **SERIES** is:

- **CONVERGENT** if

$$x_0 \leq x \leq 1;$$

- **DIVERGENT** if

$$x > 1;$$

- **INDETERMINATE** if

$$0 < x < x_0.$$

Summary scheme:

		0		x_0		1	
♯	♯	IND	C	C	C	D

We can find an approximation of x_0 which represents the solution of the equation $x^2 + \log x + 1 = 0$. We already know graphically that $0 < x_0 < 1$, and x_0 can be seen as a

4.12 Exercise 12

zero of the function $f(x) = x^2 + \log x + 1$. This function is increasing for each $x \in \mathcal{D}$, where $\mathcal{D} = (0, +\infty)$ represents its domain. We calculate the numerical value $f(1/2) = 5/4 - \log 2 \simeq 0.557 > 0$, being positive as well as $f(1) = 2 > 0$ we can conclude that $0 < x_0 < 1/2$. Continuing with the bisection method we calculate $f(1/4) = 17/16 - \log 4 \simeq -0.324 < 0$. We deduce that $1/4 < x_0 < 1/2$, we calculate $f(3/8) \simeq 0.160 > 0$ (being $3/8$ the average value between $1/4$ and $1/2$), hence $1/4 < x_0 < 3/8$ which we write as $0.25 < x_0 < 0.375$. Continuing

$$(0.25 + 0.375)/2 \simeq 0.313, \quad f(0.313) \simeq -0.064 < 0,$$

$$0.313 < x_0 < 0.375, \quad (0.313 + 0.375)/2 \simeq 0.344,$$

$$f(0.344) \simeq 0.051 > 0, \quad 0.313 < x_0 < 0.344,$$

$$(0.313 + 0.344)/2 \simeq 0.323, \quad f(0.323) \simeq -0.026 < 0,$$

$$0.323 < x_0 < 0.344.$$

You can proceed until you reach the desired approximation. A numerical calculation gives the approximate solution $x_0 \simeq 0.330$.

4.13 Exercise 13

Text

Study the behavior of the series of functions

$$\sum_{n=1}^{\infty} \frac{(x\sqrt{x})^{n+1}}{\sqrt{n}\log(n+1)}$$

Solution

In this case the existence condition applies only to the square root, from which $x \geq 0$, hence the domain is $\mathcal{D} = \{x \in \mathbb{R} : x \geq 0\}$.

We solve the inequality

$$x\sqrt{x} > 0.$$

The solution, being $\sqrt{x} \geq 0$, is

$$x > 0.$$

So the series is with **positive terms** if $x \in A$ with

$$A \equiv \{x \in \mathbb{R} : x > 0\},$$

with **alternate sign terms** if $x \in B$ with

$$B \equiv \{\},$$

in fact the quantity $x\sqrt{x}$ is never negative, and with null terms, and so **convergent**, if

$$x \in Z \equiv \{0\}.$$

In this case the series has positive terms, therefore we apply the asymptotic ratio criterion without considering the absolute series

$$\left|\frac{a_{n+1}}{a_n}\right| = \frac{(x\sqrt{x})^{n+2}}{\sqrt{n+1}\log(n+2)} \cdot \frac{\sqrt{n}\log(n+1)}{(x\sqrt{x})^{n+1}}$$

$$= x\sqrt{x} \frac{\sqrt{n}\log(n+1)}{\sqrt{n+1}\log(n+2)},$$

$$\lim_{n\to\infty}\left|\frac{a_{n+1}}{a_n}\right| = x\sqrt{x} \cdot \lim_{n\to\infty}\left(\frac{\sqrt{n}\log(n+1)}{\sqrt{n+1}\log(n+2)}\right) = x\sqrt{x}.$$

Where we used the following results

$$\lim_{n\to\infty}\frac{\sqrt{n}}{\sqrt{n+1}} = 1, \quad \lim_{n\to\infty}\frac{\log(n+1)}{\log(n+2)} = 1.$$

4.13 Exercise 13

We solve the inequality

$$x\sqrt{x} < 1,$$

remembering that we are interested in studying the behavior of the series when $x\sqrt{x} > 0$, i.e. for $x > 0$ (in fact for $x\sqrt{x} = 0$ the series is convergent, as previously determined). Under these conditions we can write

$$\sqrt{x} < \frac{1}{x},$$

from which

$$\begin{cases} x < 1/x^2 \\ x > 0 \end{cases}, \quad \begin{cases} x^3 < 1 \\ x > 0 \end{cases}.$$

The solution is

$$0 < x < 1.$$

The series is **convergent** if $x \in E$ with

$$E \equiv \{x \in \mathbb{R} : 0 < x < 1\},$$

divergent if $x \in F$ with

$$F \equiv \{x \in \mathbb{R} : x > 1\},$$

with unknown behavior if

$$x \in S \equiv \{1\}.$$

We study the behavior for $x = 1$, the series becomes

$$\sum_{n=1}^{\infty} \frac{1}{\sqrt{n}\log(n+1)}.$$

We apply the asymptotic ratio criterion with the divergent harmonic series

$$\sum_{n=2}^{\infty} \frac{1}{n},$$

calculating

$$\lim_{n \to \infty} \frac{1/[\sqrt{n}\log(n+1)]}{1/n} = \lim_{n \to \infty} \frac{n}{\sqrt{n}\log(n+1)}$$

$$= \lim_{n \to \infty} \frac{\sqrt{n}}{\log(n+1)} = \infty.$$

We conclude that for $x = 1$ the initial series is **divergent**.

Summarizing:

The **SERIES** is:

4.13 Exercise 13

- **CONVERGENT** if

$$0 \leq x < 1;$$

- **DIVERGENT** if

$$x \geq 1.$$

Summary scheme:

	0		1	
$\nexists$	C	C	D	D

4.14 Exercise 14

Text

Study the behavior of the series of functions

$$\sum_{n=1}^{\infty} \frac{\arctan n}{n} (x^3 - 1)^n$$

Solution

There are no restrictions on the values that the x variable can take, so the domain is $\mathcal{D} = \mathbb{R}$.

We solve the inequality

$$x^3 - 1 > 0.$$

The solution is

$$x > 1.$$

Since $\forall n \geq 1$

$$\frac{\arctan n}{n} > 0,$$

The series is with **positive terms** if $x \in A$ with

$$A \equiv \{x \in \mathbb{R} : x > 1\},$$

4.14 Exercise 14

with **alternate sign terms** if $x \in B$ with

$$B \equiv \{x \in \mathbb{R} : x < 1\},$$

with null terms, and therefore **convergent**, if

$$x \in Z \equiv \{1\}.$$

We consider the absolute series

$$\sum_{n=1}^{\infty} \left| \frac{\arctan n}{n} (x^3 - 1)^n \right| = \sum_{n=1}^{\infty} \frac{\arctan n}{n} |x^3 - 1|^n$$

in fact $\forall n > 1$ we have $n > 0$ and $\arctan n > 0$, and we apply the asymptotic ratio criterion

$$\left| \frac{a_{n+1}}{a_n} \right| = \frac{\arctan(n+1)}{n+1} |x^3 - 1|^{n+1} \cdot \frac{n}{\arctan n} \frac{1}{|x^3 - 1|^n}$$

$$= |x^3 - 1| \cdot \frac{\arctan(n+1)}{\arctan n} \cdot \frac{n}{n+1},$$

from which

$$\lim_{n \to \infty} \left| \frac{a_{n+1}}{a_n} \right| = |x^3 - 1| \cdot \lim_{n \to \infty} \left(\frac{\arctan(n+1)}{\arctan n} \cdot \frac{n}{n+1} \right) = |x^3 - 1|,$$

having used

$$\lim_{n \to \infty} \frac{\arctan(n+1)}{\arctan n} = 1, \quad \lim_{n \to \infty} \frac{n}{n+1} = 1.$$

We solve the inequality

$$|x^3 - 1| < 1,$$

which is equivalent to the system

$$\begin{cases} x^3 - 1 < 1 \\ x^3 - 1 > -1 \end{cases}, \quad \begin{cases} x^3 < 2 \\ x^3 > 0 \end{cases}, \quad \begin{cases} x < \sqrt[3]{2} \\ x > 0 \end{cases},$$

whose solution is

$$0 < x < \sqrt[3]{2}.$$

The series is **absolutely convergent** and so **convergent** if $x \in E$ with

$$E \equiv \{x \in \mathbb{R} : 0 < x < \sqrt[3]{2}\},$$

absolutely divergent if $x \in F$ with

$$F \equiv \{x \in \mathbb{R} : x < 0 \lor x > \sqrt[3]{2}\},$$

with unknown behavior if

$$x \in S \equiv \{0, \sqrt[3]{2}\}.$$

4.14 Exercise 14

We search the values of x for which the series is both absolutely divergent and with positive terms that can be obtained intersecting the respective sets A and F. We can calculate

$$D \equiv A \cap F = \{x \in \mathbb{R} : x > \sqrt[3]{2}\}.$$

For $x \in D$ the series is **divergent**, because the study of absolute divergence coincides with that of simple divergence, being a positive terms series.

If the series is absolutely divergent ($x \in F$) and with alternate sign terms ($x \in B$), i.e. for

$$x \in G \equiv B \cap F = \{x < 0\},$$

we can write

$$\sum_{n=1}^{\infty} \frac{\arctan n}{n} (x^3 - 1)^n = \sum_{n=1}^{\infty} \frac{(-1)^n \arctan n}{n} |x^3 - 1|^n,$$

having used, in this case,

$$x^3 - 1 = -|x^3 - 1|.$$

We can apply the Leibniz criterion, being an alternating series. Knowing that for $x \in G$ (thanks to the absolute divergence)
$$\lim_{n \to \infty} \left| \frac{a_{n+1}}{a_n} \right| > 1,$$
as previously calculated, the positive terms sequence a_n is not decreasing from a certain n onwards and the series is **indeterminate** for $x \in G$.

We determine the behavior of the initial series for the values $x \in S$ for which the asymptotic ratio criterion has not provided informations (i.e. when $|x^3 - 1| = 1$). We consider $x = 0$, in this case we have $x^3 - 1 = -1$, and the series becomes
$$\sum_{n=1}^{\infty} \frac{(-1)^n \arctan n}{n}.$$
We apply the Leibniz criterion by calculating
$$\lim_{n \to \infty} \frac{\arctan n}{n} = 0,$$
in fact
$$\lim_{n \to \infty} \arctan n = \frac{\pi}{2},$$

4.14 Exercise 14

moreover the positive terms sequence

$$b_n = \frac{\arctan n}{n}$$

is definitively not increasing and to demonstrate it, we observe that the numerator is a function (sequence) of n with a growth rate greater than the denominator.

We calculate the growth rates of both functions, assuming, in order to calculate the derivatives, that $n \in \mathbb{R}$. The growth rate of the numerator is

$$(\arctan n)' = \frac{1}{1+n^2},$$

while that of the denominator is $n' = 1$. Being

$$\frac{1}{1+n^2} < 1$$

$\forall n$, we conclude that the sequence with positive terms

$$\frac{\arctan n}{n}$$

is not increasing and consequentially for $x = 0$, the initial series is **convergent**.

Figure 4.9 shows the graph of the b_n sequence. Similarly, for $x = \sqrt[3]{2}$, where $x^3 - 1 = 1$, we have

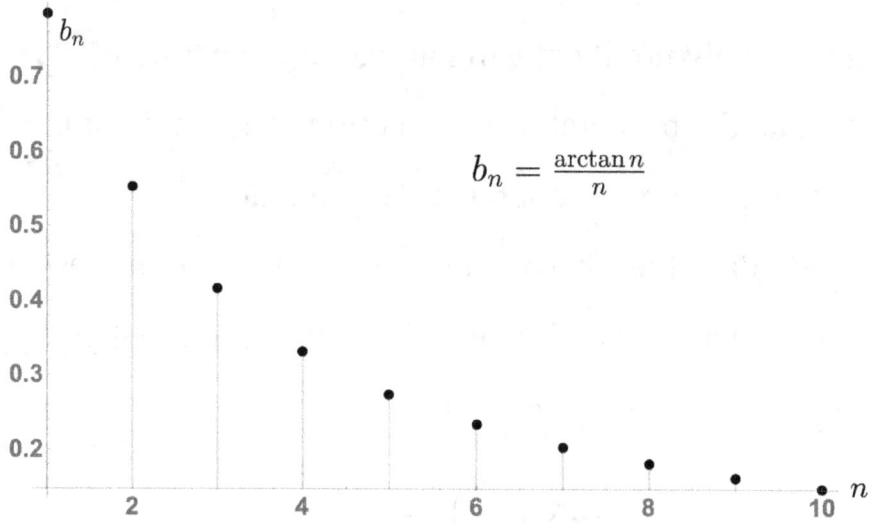

Figure 4.9: Plot of the sequence b_n for $n \in [1, 10] \subset \mathbb{R}$.

$$\sum_{n=1}^{\infty} \frac{\arctan n}{n}.$$

We apply the asymptotic ratio criterion with the divergent harmonic series

$$\sum_{n=1}^{\infty} \frac{1}{n},$$

4.14 Exercise 14

calculating

$$\lim_{n \to \infty} \frac{\arctan n / n}{1/n} = \lim_{n \to \infty} \arctan n = \frac{\pi}{2}$$

we observe that the series have the same behavior.

We conclude that for

$$x = \sqrt[3]{2}$$

the initial series is **divergent**.

Summarizing:

The **SERIES** is:

- **CONVERGENT** if

$$0 \leq x < \sqrt[3]{2};$$

- **DIVERGENT** if

$$x \geq \sqrt[3]{2};$$

- **INDETERMINATE** if

$$x < 0.$$

Summary scheme:

IND	0		$\sqrt[3]{2}$	
	C	C	D	D

4.15 Exercise 15

Text

Study the behavior of the series of functions

$$\sum_{n=2}^{\infty} \frac{(1-e^x)^n}{n\sqrt{\log n}}$$

Solution

There are no restrictions on the values that the x variable can assume, so the domain is $\mathcal{D} = \mathbb{R}$.

We solve the inequality

$$1 - e^x > 0, \quad e^x < 1, \quad x < \log 1,$$

from which

$$x < 0.$$

So the series is with **positive terms** if $x \in A$ with

$$A \equiv \{x \in \mathbb{R} : x < 0\},$$

with **alternate sign terms** if $x \in B$ with

$$B \equiv \{x \in \mathbb{R} : x > 0\},$$

with null terms and therefore **convergent**, if

$$x \in Z \equiv \{0\}.$$

We consider the absolute series

$$\sum_{n=2}^{\infty} \left| \frac{(1-e^x)^n}{n\sqrt{\log n}} \right| = \sum_{n=2}^{\infty} \frac{|1-e^x|^n}{n\sqrt{\log n}}$$

in fact $\forall n > 1$ we have $n\sqrt{\log n} > 0$, and we apply the asymptotic ratio criterion

$$\left| \frac{a_{n+1}}{a_n} \right| = \frac{|1-e^x|^{n+1}}{(n+1)\sqrt{\log(n+1)}} \cdot \frac{n\sqrt{\log n}}{|1-e^x|^n}$$

$$= |1-e^x| \cdot \frac{n\sqrt{\log n}}{(n+1)\sqrt{\log(n+1)}},$$

$$\lim_{n\to\infty} \left| \frac{a_{n+1}}{a_n} \right| = |1-e^x| \cdot \lim_{n\to\infty} \left(\frac{n\sqrt{\log n}}{(n+1)\sqrt{\log(n+1)}} \right)$$

$$= |1-e^x|.$$

We solve the inequality

$$|1-e^x| < 1,$$

4.15 Exercise 15

which is equivalent to the system

$$\begin{cases} 1 - e^x < 1 \\ 1 - e^x > -1 \end{cases}$$

The first inequality is $1 - e^x < 1$, $e^x > 0$ which is true for any real x. The second $1 - e^x > -1$, $e^x < 2$ has the solution

$$x < \log 2.$$

We can conclude that the series is **absolutely convergent** and so it is also **convergent** if $x \in E$ with

$$E \equiv \{x \in \mathbb{R} : x < \log 2\},$$

absolutely divergent if $x \in F$ with

$$F \equiv \{x \in \mathbb{R} : x > \log 2\},$$

with unknown behavior if

$$x \in S \equiv \{\log 2\}.$$

We find the values of x for which the series is both absolutely divergent and with positive terms that can be obtained intersecting the respective sets A and F. We obtain

$$D \equiv A \cap F = \{\},$$

there are no values of x for which the series is simultaneously positive and absolutely divergent.

If the series is absolutely divergent ($x \in F$) and with alternate sign terms ($x \in B$), i.e.

$$x \in G \equiv B \cap F = \{x > \log 2\},$$

we can write

$$\sum_{n=2}^{\infty} \frac{(1-e^x)^n}{n\sqrt{\log n}} = \sum_{n=2}^{\infty} (-1)^n \frac{|1-e^x|^n}{n\sqrt{\log n}},$$

having used, in this case,

$$1 - e^x = -|1 - e^x|.$$

We can apply the Leibniz criterion, being an alternating series. Knowing that for $x \in G$ (thanks to the absolute

4.15 Exercise 15

divergence)
$$\lim_{n\to\infty}\left|\frac{a_{n+1}}{a_n}\right|>1,$$
as previously calculated, the positive terms sequence a_n is not decreasing from a certain n onwards and the series is **indeterminate** for $x \in G$.

We have to determine the behavior of the initial series for the values $x \in S$ for which the asymptotic ratio criterion has provided no informations (i.e. when $|1-e^x|=1$). We put $x = \log 2$, so $1 - e^x = -1$, and the series becomes
$$\sum_{n=2}^{\infty}\frac{(-1)^n}{n\sqrt{\log n}}.$$
We apply the Leibniz criterion by calculating
$$\lim_{n\to\infty}\frac{1}{n\sqrt{\log n}}=0,$$
moreover the positive terms sequence
$$\frac{1}{n\sqrt{\log n}}$$
is not increasing, being $n\sqrt{\log n}$, in the denominator, an increasing sequence, hence, for $x = \log 2$ the series is **convergent**.

Summarizing:

The **SERIES** is:

- **CONVERGENT** if

$$x \leq \log 2;$$

- **INDETERMINATE** if

$$x > \log 2.$$

Summary scheme:

	log 2	
C	C	IND

4.16 Exercise 16

Text

Study the behavior of the series of functions

$$\sum_{n=1}^{\infty} \frac{x^{1/n}}{\sqrt{n+x^2+3}}$$

Solution

The base of the power function must be positive or null, because the exponent $n \log n$ assumes non-integer values. We observe that the denominator never vanishes and that, in particular, the radicand in the denominator is always positive, $n + x^2 + 3 > 0$, being the sum of three terms that are always positive. The only condition is the following

$$x \geq 0.$$

Therefore the domain is

$$\mathcal{D} = \{x \in \mathbb{R} : x \geq 0\}.$$

The series has positive terms, but for $x = 0$ has null terms and then is **convergent**. For the remaining cases we apply the asymptotic ratio criterion

$$\left|\frac{a_{n+1}}{a_n}\right| = \frac{x^{1/(n+1)}}{\sqrt{n+x^2+4}} \cdot \frac{\sqrt{n+x^2+3}}{x^{1/n}}$$

$$= x^{1/(n+1)-1/n} \frac{\sqrt{n+x^2+3}}{\sqrt{n+x^2+4}},$$

knowing that

$$\frac{1}{n+1} - \frac{1}{n} = \frac{n-(n+1)}{n(n+1)} = -\frac{1}{n(n+1)},$$

$$\lim_{n \to \infty} \frac{\sqrt{n+x^2+3}}{\sqrt{n+x^2+4}} = 1,$$

we obtain

$$\lim_{n \to \infty} \left|\frac{a_{n+1}}{a_n}\right| = \lim_{n \to \infty} x^{-\frac{1}{n(n+1)}} = 1.$$

In this case, the asymptotic ratio criterion does not provide further information. We apply the asymptotic ratio criterion with the generalized harmonic series

$$\sum_{n=1}^{\infty} \frac{1}{n^\alpha}.$$

4.16 Exercise 16

We have, remembering that we are in the case where $x > 0$,

$$\lim_{n \to \infty} \left| \frac{x^{1/n}}{\sqrt{n + x^2 + 3}} \right| \cdot n^\alpha = \lim_{n \to \infty} |x|^{1/n} n^{\alpha - 1/2}$$

$$= \begin{cases} \infty & \text{if } \alpha > 1/2 \\ 0 & \text{if } \alpha \leq 1/2 \end{cases}.$$

For example, we can choose, $\alpha = 2/3$. In this case the generalized harmonic series diverges and we have

$$\lim_{n \to \infty} |x|^{1/n} n^{\alpha - 1/2} = \infty.$$

We conclude that the series is **divergent** for each $x > 0$.

Summary scheme:

	0	
$\nexists$	C	D

4.17 Exercise 17

Text

Study the behavior of the series of functions

$$\sum_{n=2}^{\infty} \frac{(x^2 - 2x - 3)^{2n}}{n! e^n}$$

Solution

There are no restrictions on the values that the x variable can assume, so the domain is $\mathcal{D} = \mathbb{R}$.

We consider the power in numerator. The exponent $2n$ is always even, so the series has positive terms. We solve the equation

$$x^2 - 2x - 3 = 0, \quad \Delta = 4 + 12 = 16, \quad x_{1,2} = \frac{2 \pm 4}{2},$$

$$x_1 = -1, x_2 = 3.$$

The series has null terms, and therefore **convergent**, if

$$x \in Z \equiv \{-1, 3\}.$$

4.17 Exercise 17

We apply the asymptotic ratio criterion

$$\left|\frac{a_{n+1}}{a_n}\right| = \frac{(x^2-2x-3)^{2^{n+1}}}{(n+1)!\,e^{n+1}} \cdot \frac{n!\,e^n}{(x^2-2x-3)^{2^n}}$$

$$= \frac{1}{e} \cdot \frac{(x^2-2x-3)^{2^{n+1}-2^n}}{(n+1)},$$

$$\lim_{n\to\infty}\left|\frac{a_{n+1}}{a_n}\right| = \frac{1}{e}\lim_{n\to\infty}\frac{(x^2-2x-3)^{2^n}}{(n+1)}$$

$$= \begin{cases} \infty & \text{if } |x^2-2x-3| > 1 \\ 0 & \text{if } |x^2-2x-3| \leq 1 \end{cases}.$$

We solve the inequality

$$|x^2 - 2x - 3| < 1,$$

which is equivalent to the system

$$\begin{cases} x^2 - 2x - 3 < 1 \\ x^2 - 2x - 3 > -1 \end{cases}$$

The first inequality is $x^2 - 2x - 3 < 1$, $x^2 - 2x - 4 < 0$, we solve the associated equation that has the solutions $x^2 -$

$2x-4=0$, $\Delta=4+16=20$, $x_{1,2}=(2\pm 2\sqrt{5})/2=1\pm\sqrt{5}$, from which

$$1-\sqrt{5}<x<1+\sqrt{5},$$

while for the second $x^2-2x-3>-1$, $x^2-2x-2>0$, $x^2-2x-2=0$, $\Delta=4+8=12$, $x_{1,2}=(2\pm 2\sqrt{3})/2=1\pm\sqrt{3}$, from which

$$x<1-\sqrt{3} \ \lor \ x>1+\sqrt{3}.$$

The system becomes

$$\begin{cases} 1-\sqrt{5}<x<1+\sqrt{5} \\ x<1-\sqrt{3} \ \lor \ x>1+\sqrt{3} \end{cases},$$

whose solution is

$$1-\sqrt{5}<x<1-\sqrt{3} \ \lor \ 1+\sqrt{3}<x<1+\sqrt{5}.$$

The series is **convergent** if $x \in E$ with

$$E \equiv \{x \in \mathbb{R} : 1-\sqrt{5} \leq x \leq 1-\sqrt{3} \ \lor \ 1+\sqrt{3} \leq x \leq 1+\sqrt{5}\},$$

4.17 Exercise 17

and **divergent** if $x \in F$ with

$$F \equiv \{x \in \mathbb{R} : x < 1 - \sqrt{5} \vee 1 - \sqrt{3} < x < 1 + \sqrt{3} \\ \vee \; x > 1 + \sqrt{5}\}.$$

Summarizing:

The **SERIES** is:

- **CONVERGENT** if

$$1 - \sqrt{5} \leq x \leq 1 - \sqrt{3} \;\vee\; 1 + \sqrt{3} \leq x \leq 1 + \sqrt{5};$$

- **DIVERGENT** if

$$x < 1 - \sqrt{5} \;\vee\; 1 - \sqrt{3} < x < 1 + \sqrt{3} \;\vee\; x > 1 + \sqrt{5}.$$

Summary scheme:

	$1-\sqrt{5}$		$1-\sqrt{3}$		$1+\sqrt{3}$		$1+\sqrt{5}$	
D	C	C	C	D	C	C	C	D

4.18 Exercise 18

Text

Study the behavior of the series of functions

$$\sum_{n=1}^{\infty} \frac{1}{n^3} \left(\frac{x}{\log x} \right)^{(n+2)^2}$$

Solution

The domain is $\mathcal{D} = \{x \in \mathbb{R} : x > 0 \wedge x \neq 1\}$, in fact due to the presence of the term $\log x$, we must put $x > 0$ and due to the denominator we also have the condition $\log x \neq 0$, hence $x \neq 1$.

We observe that $(n+2)^2$ has the same parity as $n+2$ which has the same parity as n.

We solve the inequality

$$\frac{x}{\log x} > 0.$$

The numerator is positive when $x > 0$, while the denominator is positive if $\log x > 0$, i.e. if $x > 1$. Combining the results, remembering the domain, we obtain the solution

4.18 Exercise 18

of the inequality

$$x > 1.$$

So the series is with **positive terms** if $x \in A$ with

$$A \equiv \{x \in \mathbb{R} : x > 1\},$$

with **alternate sign terms** if $x \in B$ with

$$B \equiv \{x \in \mathbb{R} : 0 < x < 1\}.$$

We consider the absolute series

$$\sum_{n=1}^{\infty} \left| \frac{1}{n^3} \left(\frac{x}{\log x} \right)^{(n+2)^2} \right| = \sum_{n=1}^{\infty} \frac{1}{n^3} \left| \frac{x}{\log x} \right|^{(n+2)^2}$$

in fact $\forall n > 1$ we have $n^3 > 0$, and we apply the asymptotic ratio criterion

$$\left| \frac{a_{n+1}}{a_n} \right| = \frac{1}{(n+1)^3} \left| \frac{x}{\log x} \right|^{(n+3)^2} \cdot n^3 \left| \frac{\log x}{x} \right|^{(n+2)^2},$$

$$\lim_{n\to\infty}\left|\frac{a_{n+1}}{a_n}\right| = \lim_{n\to\infty}\frac{n^3}{(n+1)^3}\left|\frac{x}{\log x}\right|^{2n+5} = \lim_{n\to\infty}\left|\frac{x}{\log x}\right|^{2n+5}$$

$$= \begin{cases} 0 & \text{if } |x/\log x| < 1 \\ \infty & \text{if } |x/\log x| > 1 \\ 1 & \text{if } |x/\log x| = 1 \end{cases}.$$

We solve the inequality

$$\left|\frac{x}{\log x}\right| < 1,$$

which is equivalent to the system

$$\begin{cases} \frac{x}{\log x} < 1 \\ \frac{x}{\log x} > -1 \end{cases}$$

We consider the first inequality

$$\frac{x}{\log x} < 1, \quad \frac{x - \log x}{\log x} < 0,$$

The numerator is positive if $x - \log x > 0$, $x > \log x$, true for all $x \in \mathcal{D}$, as can be seen in Figure 4.10. The denominator

4.18 Exercise 18

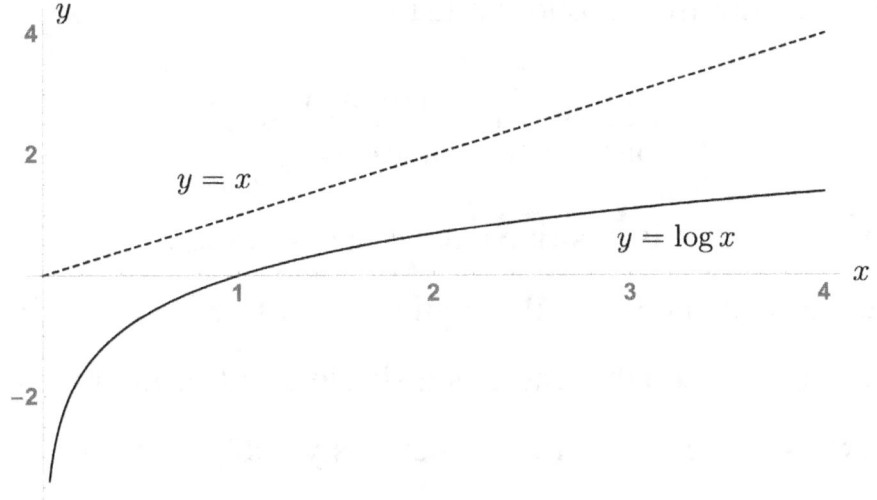

Figure 4.10: Plot of the functions $y = \log x$ (continuous line) and $y = x$ (dashed line).

is positive if $\log x > 0$, i.e. for $x > 1$, combining the results we find the solution of the first inequality of the system

$$0 < x < 1.$$

The system becomes

$$\begin{cases} 0 < x < 1 \\ \frac{x}{\log x} > -1 \end{cases}$$

We consider the second inequality

$$\frac{x}{\log x} > -1, \quad \frac{x+\log x}{\log x} > 0,$$

The numerator is positive if $x + \log x > 0$, $\log x > -x$. This inequality can be solved graphically, with reference to Figure 4.11, we see that there is a single point of intersection between the graphs of the functions $y = \log x$ and $y = -x$, whose abscissa, called x_1, is between 0 and 1 ($0 < x_1 < 1$). The solution of the inequality $\log x > -x$ can be written as $x > x_1$. The denominator, as in the other case, is positive if $\log x > 0$, i.e. for $x > 1$ and, combining the results, we find the following solution for the second inequality of the system

$$0 < x < x_1 \quad \vee \quad x > 1.$$

The system finally becomes

$$\begin{cases} 0 < x < 1 \\ 0 < x < x_1 \quad \vee \quad x > 1 \end{cases}$$

4.18 Exercise 18

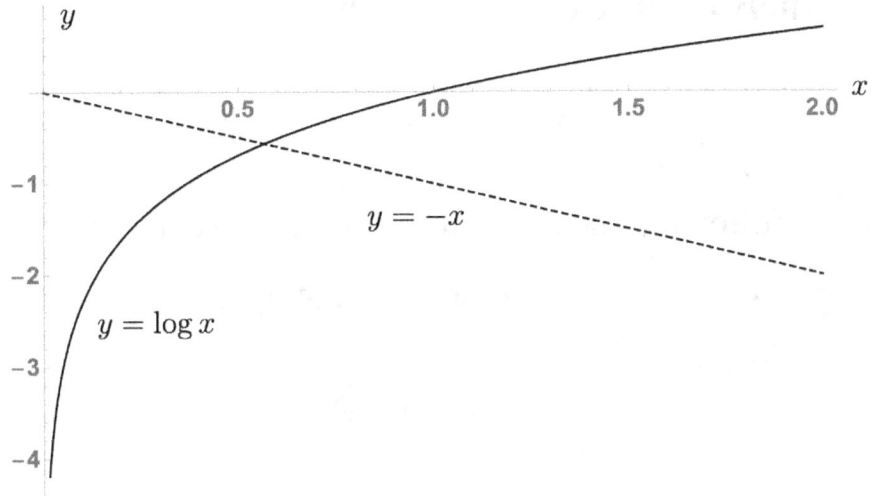

Figure 4.11: Plot of the functions $y = \log x$ (continuous line) and $y = -x$ (dashed line).

whose solution is

$$0 < x < x_1.$$

The series is **absolutely convergent** and then it is also **convergent** if $x \in E$ with

$$E \equiv \{x \in \mathbb{R} : 0 < x < x_1\},$$

absolutely divergent if $x \in F$ with

$$F \equiv \{x \in \mathbb{R} : x > x_1, x \neq 1\},$$

where, following the domain, we have also added the condition $x \neq 1$, and with unknown behavior if

$$x \in S \equiv \{x_1\}.$$

We find the values of x for which the series is both absolutely divergent and with positive terms that can be obtained intersecting the respective sets A and F. We obtain

$$D \equiv A \cap F = \{x \in \mathbb{R} : x > 1\}.$$

For $x \in D$ the series is **divergent**, because the study of absolute divergence coincides with that of simple divergence, being a positive terms series.

If the series is absolutely divergent ($x \in F$) and with alternate sign terms ($x \in B$), i.e. for

$$x \in G \equiv B \cap F = \{x_1 < x < 1\},$$

4.18 Exercise 18

we can write

$$\sum_{n=1}^{\infty} \frac{1}{n^3}\left(\frac{x}{\log x}\right)^{(n+2)^2} = \sum_{n=1}^{\infty}(-1)^n \frac{1}{n^3}\left|\frac{x}{\log x}\right|^{(n+2)^2},$$

having used, in this case,

$$\frac{x}{\log x} = -\left|\frac{x}{\log x}\right|,$$

together with the identity $(-1)^{(n+2)^2} = (-1)^n$. We can apply the Leibniz criterion, being a series with alternate sign terms. Knowing that for $x \in G$ (thanks to the absolute divergence)

$$\lim_{n \to \infty}\left|\frac{a_{n+1}}{a_n}\right| > 1,$$

as previously calculated, the positive term series a_n is not decreasing from a certain n onwards and then the series is **indeterminate** for $x \in G$.

We have to determine the behavior of the initial series for that value of $x \in S$ for which the asymptotic ratio criterion did not provide informations (i.e. when $|x/\log x| = 1$). We put $x = x_1$, in this case we have $x/\log x = -1$, and the

series becomes
$$\sum_{n=1}^{\infty}(-1)^n \frac{1}{n^3}.$$
We apply the Leibniz criterion by calculating
$$\lim_{n\to\infty}\frac{1}{n^3}=0,$$
moreover the positive terms sequence
$$\frac{1}{n^3}$$
is not increasing, being n^3, in the denominator, an increasing sequence, hence, for $x=x_1$ the series is **convergent**.

Summarizing:

The **SERIES** is:

- **CONVERGENT** if
$$0<x\leq x_1;$$

- **DIVERGENT** if
$$x>1;$$

- **INDETERMINATE** if
$$x_1<x<1.$$

4.18 Exercise 18

Summary scheme:

	0		x_1		1	
$\nexists$	$\nexists$	C	C	IND	$\nexists$	D

4.19 Exercise 19

Text

Study the behavior of the series of functions

$$\sum_{n=2}^{\infty} \frac{(4x^2-1)^n}{\log(n!)\arctan(e^n)}$$

Solution

There are no limitations on the values that the x variable can assume, so the domain is $\mathcal{D} = \mathbb{R}$.

We solve the inequality

$$4x^2 - 1 > 0.$$

The associated equation $4x^2 - 1 = 0$ has the roots $x = \pm 1/2$, hence

$$x < -\frac{1}{2} \ \lor \ x > \frac{1}{2}.$$

So the series is with **positive terms** if $x \in A$ with

$$A \equiv \left\{ x \in \mathbb{R} : x < -\frac{1}{2} \ \lor \ x > \frac{1}{2} \right\},$$

4.19 Exercise 19

with **alternate sign terms** if $x \in B$ with

$$B \equiv \left\{ x \in \mathbb{R} : -\frac{1}{2} < x < \frac{1}{2} \right\},$$

with null terms, and therefore **convergent**, if

$$x \in Z \equiv \left\{ \pm \frac{1}{2} \right\}.$$

We consider the absolute series

$$\sum_{n=2}^{\infty} \left| \frac{(4x^2 - 1)^n}{\log(n!) \arctan(e^n)} \right| = \sum_{n=2}^{\infty} \frac{|4x^2 - 1|^n}{\log(n!) \arctan(e^n)},$$

in fact $\forall n > 1$ we have $\log(n!) > 0$ and $\arctan(e^n) > 0$. We apply the asymptotic ratio criterion

$$\left| \frac{a_{n+1}}{a_n} \right| = \frac{|4x^2 - 1|^{n+1}}{\log[(n+1)!] \arctan(e^{n+1})} \cdot \frac{\log(n!) \arctan(e^n)}{|4x^2 - 1|^n}$$

$$= \frac{|4x^2 - 1| \log(n!) \arctan(e^n)}{\log[(n+1)!] \arctan(e^{n+1})},$$

$$\lim_{n \to \infty} \left| \frac{a_{n+1}}{a_n} \right| = |4x^2 - 1| \cdot \lim_{n \to \infty} \left(\frac{\log(n!)}{\log[(n+1)!]} \right),$$

in fact

$$\lim_{n \to \infty} \frac{\arctan(e^n)}{\arctan(e^{n+1})} = 1.$$

To calculate the limit we write

$$\lim_{n\to\infty} \frac{\log(n!)}{\log[(n+1)!]} = \lim_{n\to\infty} \frac{\log(n!)}{\log[(n+1)n!]}$$
$$= \lim_{n\to\infty} \frac{\log(n!)}{\log(n+1)+\log(n!)}$$
$$= \lim_{n\to\infty} \left(1+\frac{\log(n+1)}{\log(n!)}\right)^{-1} = 1,$$

in fact, using the following relation for the natural logarithm of n factorial

$$n\log n - n + 1 \leq \log(n!) \leq n\log n - n + 1 + \log n,$$

we can write

$$1 + \frac{\log(n+1)}{n(\log n - 1) + 1 + \log n} \leq 1 + \frac{\log(n+1)}{\log(n!)}$$
$$\leq 1 + \frac{\log(n+1)}{n(\log n - 1) + 1}$$

and, knowing that

$$\lim_{n\to\infty} \left(1 + \frac{\log(n+1)}{n(\log n - 1) + 1 + \log n}\right) = 1,$$

$$\lim_{n\to\infty} \left(1 + \frac{\log(n+1)}{n(\log n - 1) + 1}\right) = 1,$$

4.19 Exercise 19

we conclude, using the ratio criterion, that also

$$\lim_{n \to \infty} \left(1 + \frac{\log(n+1)}{\log(n!)}\right) = 1, \quad \lim_{n \to \infty} \left(1 + \frac{\log(n+1)}{\log(n!)}\right)^{-1} = 1.$$

We apply the asymptotic ratio criterion, by calculating

$$\lim_{n \to \infty} \left|\frac{a_{n+1}}{a_n}\right| = |4x^2 - 1|,$$

we solve the inequality

$$|4x^2 - 1| < 1,$$

which is equivalent to the system

$$\begin{cases} 4x^2 - 1 < 1 \\ 4x^2 - 1 > -1 \end{cases}$$

The first inequality is $4x^2 - 1 < 1$, $4x^2 - 2 < 0$ (the associated equation $4x^2 - 2 = 0$ has solutions $x_{1,2} = \pm\sqrt{2}/2$) and has the solution

$$\frac{-\sqrt{2}}{2} < x < \frac{\sqrt{2}}{2},$$

while the second $4x^2 - 1 > -1$, $4x^2 > 0$ has the solution

$$x \neq 0.$$

The system becomes

$$\begin{cases} -\frac{\sqrt{2}}{2} < x < \frac{\sqrt{2}}{2} \\ x \neq 0 \end{cases},$$

whose solution is

$$-\frac{\sqrt{2}}{2} < x < 0 \quad \vee \quad 0 < x < \frac{\sqrt{2}}{2}.$$

The series is **absolutely convergent** and then it is also **convergent** if $x \in E$ with

$$E \equiv \left\{ x \in \mathbb{R} : -\frac{\sqrt{2}}{2} < x < \frac{\sqrt{2}}{2}, x \neq 0 \right\},$$

absolutely divergent if $x \in F$ with

$$F \equiv \left\{ x \in \mathbb{R} : x < -\frac{\sqrt{2}}{2} \vee x > \frac{\sqrt{2}}{2} \right\},$$

with unknown behavior if

$$x \in S \equiv \left\{ 0, \pm \frac{\sqrt{2}}{2} \right\}.$$

4.19 Exercise 19

We search the values of x for which the series is both absolutely divergent and with positive terms that can be obtained intersecting the respective sets A and F. We obtain

$$D \equiv A \cap F = \left\{ x \in \mathbb{R} : x < -\frac{\sqrt{2}}{2} \lor x > \frac{\sqrt{2}}{2} \right\}.$$

For $x \in D$ the series is **divergent**, because the study of absolute divergence coincides with that of simple divergence, being a positive terms series.

The series is never simultaneously with alternate sign terms ($x \in B$) and absolutely divergent ($x \in F$), in fact

$$G \equiv B \cap F = \{\}.$$

We determine the behavior of the initial series for the values $x \in S$ for which the asymptotic ratio criterion has provided no informations (i.e. when $|4x^2 - 1| = 1$). We put $x = 0$, in this case we have $4x^2 - 1 = -1$, and the series becomes

$$\sum_{n=2}^{\infty} \frac{(-1)^n}{\log(n!) \arctan(e^n)}.$$

We apply the Leibniz criterion by calculating

$$\lim_{n \to \infty} \frac{1}{\log(n!) \arctan(e^n)} = 0,$$

moreover the positive terms sequence

$$\frac{1}{\log(n!) \arctan(e^n)}$$

is not increasing, being the reciprocal of an increasing sequence, therefore, for $x = 0$ the series is **convergent**. Similarly, for $x = \pm\sqrt{2}/2$, where $4x^2 - 1 = 1$, we have

$$\sum_{n=2}^{\infty} \frac{1}{\log(n!) \arctan(e^n)}.$$

We have

$$\frac{1}{\log(n!) \arctan(e^n)} > \frac{2}{\pi \log(n^n)} = \frac{2}{\pi} \frac{1}{n \log n},$$

hence, being the generalized harmonic series of the second type

$$\sum_{n=2}^{\infty} \frac{1}{n \log(n)}$$

divergent, we obtain from the ratio criterion that for $x = \pm\sqrt{2}/2$, the series is **divergent**.

4.19 Exercise 19

Summarizing:

The **SERIES** is:

- **CONVERGENT** if

$$-\frac{\sqrt{2}}{2} < x < \frac{\sqrt{2}}{2};$$

- **DIVERGENT** if

$$x \leq -\frac{\sqrt{2}}{2} \lor x \geq \frac{\sqrt{2}}{2}.$$

Summary scheme:

	$-\sqrt{2}/2$		$\sqrt{2}/2$	
D	D	C	D	D

4.20 Exercise 20

Text

Study the behavior of the series of functions

$$\sum_{n=2}^{\infty} \frac{(8-x^2)^{n+1/n}}{\sqrt{n^3-1}}$$

Solution

The base of the power in the numerator must be non-negative, due to the exponent, $n+1/n$, which assumes non-integer values. To find the domain we need to solve the following inequality

$$8-x^2 \geq 0,$$

The associated equation has the solutions $x_{1,2} = \pm\sqrt{8} = \pm 2\sqrt{2}$, hence the solution of the inequality

$$-2\sqrt{2} \leq x \leq 2\sqrt{2}.$$

The domain is

$$\mathcal{D} = \{x \in \mathbb{R} : -2\sqrt{2} \leq x \leq 2\sqrt{2}\}.$$

4.20 Exercise 20

For $x = \pm 2\sqrt{2}$ the series has null terms and so it is **convergent**. For all the other values of x (admitted by the domain) the series has positive terms. We apply the asymptotic ratio criterion

$$\left|\frac{a_{n+1}}{a_n}\right| = \frac{(8-x^2)^{n+1+1/(n+1)}}{\sqrt{(n+1)^3-1}} \cdot \frac{\sqrt{n^3-1}}{(8-x^2)^{n+1/n}}$$

$$= \frac{\sqrt{n^3-1}}{\sqrt{(n+1)^3-1}} \cdot (8-x^2)^{1+1/(n+1)-1/n},$$

$$\lim_{n\to\infty} \left|\frac{a_{n+1}}{a_n}\right| = \lim_{n\to\infty} \frac{\sqrt{n^3-1}}{\sqrt{(n+1)^3-1}} \cdot (8-x^2)^{1-1/(n^2+n)}.$$

We have

$$\lim_{n\to\infty} \frac{\sqrt{n^3-1}}{\sqrt{(n+1)^3-1}} = 1,$$

moreover

$$\lim_{n\to\infty} (8-x^2)^{1-1/(n^2+n)} = 8-x^2,$$

so

$$\lim_{n\to\infty} \left|\frac{a_{n+1}}{a_n}\right| = 8-x^2.$$

We solve the inequality

$$8-x^2 < 1, \quad 7-x^2 < 0,$$

the associated equation has the solutions $x_{1,2} = \pm\sqrt{7}$, hence the solution

$$-2\sqrt{2} \leq x < -\sqrt{7} \vee \sqrt{7} < x \leq 2\sqrt{2},$$

where we have also included the conditions imposed by the domain.

The series is **convergent** if $x \in E$ with

$$E \equiv \{x \in \mathbb{R} : -2\sqrt{2} \leq x < -\sqrt{7} \vee \sqrt{7} < x \leq 2\sqrt{2}\},$$

and **divergent** if $x \in F$ with

$$F \equiv \{x \in \mathbb{R} : -\sqrt{7} < x < \sqrt{7}\},$$

with unknown behavior if

$$x \in S \equiv \{\pm\sqrt{7}\}.$$

We study the latter case, for $x = \pm\sqrt{7}$ the initial series becomes

$$\sum_{n=2}^{\infty} \frac{1}{\sqrt{n^3 - 1}},$$

4.20 Exercise 20

We use the asymptotic ratio criterion with the convergent generalized harmonic series

$$\sum_{n=2}^{\infty} \frac{1}{n^{3/2}},$$

we calculate

$$\lim_{n \to \infty} \frac{1/\sqrt{n^3-1}}{1/n^{3/2}} = \lim_{n \to \infty} \frac{n^{3/2}}{(n^3-1)^{1/2}}$$

$$= \lim_{n \to \infty} \frac{n^{3/2}}{\left(n^3(1-1/n^3)\right)^{1/2}} = 1,$$

we observe that the series have the same behavior and so the initial series, for $x = \pm\sqrt{7}$, is **convergent**.

The **SERIES** is:

- **CONVERGENT** if

$$-2\sqrt{2} \leq x \leq -\sqrt{7} \ \lor \ \sqrt{7} \leq x \leq 2\sqrt{2};$$

- **DIVERGENT** if

$$-\sqrt{7} < x < \sqrt{7}.$$

Summary scheme:

♯	$-2\sqrt{2}$		$-\sqrt{7}$		$\sqrt{7}$		$2\sqrt{2}$	♯
	C	C	C	D	C	C	C	

4.21 Exercise 21

Text

Study the behavior of the series of functions

$$\sum_{n=1}^{\infty} \frac{(8-x^2)^n}{n^2+3n-1}$$

Solution

There are no limitations on the values that the x variable can assume, so the domain is simply $\mathcal{D} = \mathbb{R}$.

We solve the inequality

$$8 - x^2 > 0.$$

The associated equation $8 - x^2 = 0$ has the roots $x = \pm\sqrt{8} = \pm 2\sqrt{2}$, hence

$$-2\sqrt{2} < x < 2\sqrt{2}.$$

So the series is with **positive terms** if $x \in A$ with

$$A \equiv \{x \in \mathbb{R} : -2\sqrt{2} < x < 2\sqrt{2}\},$$

with **alternate sign terms** if $x \in B$ with

$$B \equiv \{x \in \mathbb{R} : x < -2\sqrt{2} \lor x > 2\sqrt{2}\},$$

with null terms, and therefore **convergent**, if

$$x \in Z \equiv \{\pm 2\sqrt{2}\}.$$

We consider the absolute series

$$\sum_{n=1}^{\infty} \left| \frac{(8-x^2)^n}{n^2+3n-1} \right| = \sum_{n=1}^{\infty} \frac{|8-x^2|^n}{n^2+3n-1}$$

in fact $\forall n > 1$ we have $n^2 + 3n - 1 > 0$, and we apply the asymptotic ratio criterion

$$\left| \frac{a_{n+1}}{a_n} \right| = \frac{|8-x^2|^{n+1}}{(n+1)^2+3(n+1)-1} \cdot \frac{n^2+3n-1}{|8-x^2|^n}$$

$$= |8-x^2| \cdot \frac{n^2+3n-1}{n^2+5n+3},$$

$$\lim_{n \to \infty} \left| \frac{a_{n+1}}{a_n} \right| = |8-x^2| \cdot \lim_{n \to \infty} \left(\frac{n^2+3n-1}{n^2+5n+3} \right) = |8-x^2|,$$

in fact

$$\lim_{n \to \infty} \frac{n^2+3n-1}{n^2+5n+3} = 1,$$

being the ratio of polynomials of the same degree.

We solve the inequality

$$|8-x^2| < 1,$$

4.21 Exercise 21

which is equivalent to the system

$$\begin{cases} 8 - x^2 < 1 \\ 8 - x^2 > -1 \end{cases}$$

The first inequality is $8 - x^2 < 1$, $x^2 > 7$ and has the solution

$$x < -\sqrt{7} \ \lor \ x > \sqrt{7},$$

while the second one is $8 - x^2 > -1$, $x^2 < 9$ and has the solution

$$-3 < x < 3.$$

The system becomes

$$\begin{cases} x < -\sqrt{7} \ \lor \ x > \sqrt{7} \\ -3 < x < 3 \end{cases},$$

whose solution is

$$-3 < x < -\sqrt{7} \ \lor \ \sqrt{7} < x < 3.$$

The series is **absolutely convergent** and so it is also **convergent** if $x \in E$ with

$$E \equiv \{x \in \mathbb{R} : -3 < x < -\sqrt{7} \vee \sqrt{7} < x < 3\},$$

absolutely divergent if $x \in F$ with

$$F \equiv \{x \in \mathbb{R} : x < -3 \vee -\sqrt{7} < x < \sqrt{7} \vee x > 3\},$$

with unknown behavior if

$$x \in S \equiv \{\pm\sqrt{7}, \pm 3\}.$$

We search the values of x for which the series is both absolutely divergent and with positive terms that can be obtained intersecting the respective sets A and F. We obtain

$$D \equiv A \cap F = \{x \in \mathbb{R} : -\sqrt{7} < x < \sqrt{7}\}.$$

For $x \in D$ the series is **divergent**, because the study of absolute divergence coincides with that of simple divergence, being a positive terms series.

4.21 Exercise 21

If the series is absolutely divergent ($x \in F$) and with alternate sign terms ($x \in B$), i.e. for

$$x \in G \equiv B \cap F = \{x < -3 \vee x > 3\},$$

we can write

$$\sum_{n=1}^{\infty} \frac{(8-x^2)^n}{n^2 + 3n - 1} = \sum_{n=1}^{\infty} \frac{(-1)^n |8-x^2|^n}{n^2 + 3n - 1},$$

having used, in this case,

$$8 - x^2 = -|8 - x^2|.$$

We can apply the Leibniz criterion, being a series with alternate sign terms. Knowing that for $x \in G$ (thanks to the absolute divergence)

$$\lim_{n \to \infty} \left| \frac{a_{n+1}}{a_n} \right| > 1,$$

as previously calculated, the positive terms sequence a_n is not decreasing from a certain n onwards and so the series is **indeterminate** for $x \in G$.

We determine the behavior of the initial series for the values $x \in S$ for which the asymptotic ratio criterion has provided no informations (i.e. when $|8 - x^2| = 1$). We consider $x = \pm 3$, in this case we have $8 - x^2 = -1$, and the series becomes

$$\sum_{n=1}^{\infty} \frac{(-1)^n}{n^2 + 3n - 1}.$$

We apply the Leibniz criterion by calculating

$$\lim_{n \to \infty} \frac{1}{n^2 + 3n - 1} = 0,$$

moreover the positive terms sequence

$$\frac{1}{n^2 + 3n - 1}$$

is not increasing, being $n^2 + 3n - 1$, in the denominator, an increasing sequence, and therefore, for $x = \pm 3$ the series is **convergent**. Similarly, for $x = \pm\sqrt{7}$, where $8 - x^2 = 1$, we have

$$\sum_{n=1}^{\infty} \frac{1}{n^2 + 3n - 1}.$$

4.21 Exercise 21

We apply the ratio criterion, in fact we can choose a $\alpha \in \mathbb{R}^+$ such that
$$\frac{1}{n^2+3n-1} < \frac{\alpha}{n^2},$$
from a certain n onwards, in this way the series will be convergent, being the generalized harmonic series
$$\sum_{n=1}^{\infty} \frac{1}{n^2}$$
convergent. We solve
$$n^2 < (n^2+3n-1)\alpha, \quad \alpha > \frac{n^2}{n^2+3n-1} = \frac{1}{1+3/n-1/n^2}.$$
Being
$$\frac{1}{1+3/n-1/n^2} < 1,$$
in fact
$$\frac{3}{n}-\frac{1}{n^2} > 0, \quad \frac{3}{n} > \frac{1}{n^2}, \quad 3n > 1, \quad n > \frac{1}{3},$$
because $n > 1$, we can simply choose $\alpha = 1$ and conclude that since
$$\frac{1}{n^2+3n-1} < \frac{1}{n^2},$$

for $x = \pm\sqrt{7}$, the series is **convergent**.

Summarizing:

The **SERIES** is:

- **CONVERGENT** if

$$-3 \leq x \leq -\sqrt{7} \ \vee \ \sqrt{7} \leq x \leq 3;$$

- **DIVERGENT** if

$$-\sqrt{7} < x < \sqrt{7};$$

- **INDETERMINATE** if

$$x < -3 \ \vee \ x > 3.$$

Summary scheme:

	-3		$-\sqrt{7}$		$\sqrt{7}$		3	
IND	C	C	C	D	C	C	C	IND

4.22 Exercise 22

Text

Study the behavior of the series of functions

$$\sum_{n=1}^{\infty} \frac{(x(\log x + 2))^{n^2}}{n \arctan n}$$

Solution

Given the presence of $\log x$ the existence condition is $x > 0$. The domain is simply $\mathcal{D} = \mathbb{R}^+$.

Having n^2 the same parity as n, we solve the inequality

$$x(\log x + 2) > 0.$$

The first factor, x, is positive if $x > 0$, the second, $\log x + 2$, is positive if $\log x > -2$, i.e. if $x > e^{-2}$. By combining the results and considering the domain we obtain that the product is positive if

$$x > e^{-2}.$$

The series is with **positive terms** if $x \in A$ with

$$A \equiv \{x \in \mathbb{R} : x > e^{-2}\};$$

with **alternate sign terms** if $x \in B$ with

$$B \equiv \{x \in \mathbb{R} : 0 < x < e^{-2}\};$$

with null terms, and therefore **convergent**, if

$$x \in Z \equiv \{e^{-2}\}.$$

We consider the absolute series

$$\sum_{n=1}^{\infty} \left| \frac{(x(\log x + 2))^{n^2}}{n \arctan n} \right| = \sum_{n=1}^{\infty} \frac{|x(\log x + 2)|^{n^2}}{n \arctan n}$$

in fact $\forall n > 1$ we have $n \arctan n > 0$, and we apply the asymptotic ratio criterion

$$\left| \frac{a_{n+1}}{a_n} \right| = \frac{|x(\log x + 2)|^{(n+1)^2}}{(n+1) \arctan(n+1)} \cdot \frac{n \arctan n}{|x(\log x + 2)|^{n^2}},$$

hence, being

$$\lim_{n \to \infty} \frac{n \arctan n}{(n+1) \arctan(n+1)} \lim_{n \to \infty} \frac{n \cdot (\pi/2)}{(n+1) \cdot (\pi/2)} = 1,$$

4.22 Exercise 22

$$\lim_{n \to \infty} \left| \frac{a_{n+1}}{a_n} \right| = \lim_{n \to \infty} |x(\log x + 2)|^{2n+1}$$

$$= \begin{cases} 0 & \text{if } |x(\log x + 2)| < 1 \\ \infty & \text{if } |x(\log x + 2)| > 1 \\ 1 & \text{if } |x(\log x + 2)| = 1 \end{cases}.$$

We solve the inequality

$$|x(\log x + 2)| < 1,$$

which is equivalent to the system

$$\begin{cases} x(\log x + 2) < 1 \\ x(\log x + 2) > -1 \end{cases}, \quad \begin{cases} x \log x < 1 - 2x \\ x \log x > -1 - 2x \end{cases},$$

$$\begin{cases} \log x < 1/x - 2 \\ \log x > -1/x - 2 \end{cases},$$

in fact, thanks to the domain, $x > 0$.

We consider the first inequality $\log x < 1/x - 2$, since we cannot solve it analytically, we proceed graphically, drawing the plot of the function $y = \log x$ and that of the function $y = 1/x - 2$, the latter being a translated hyperbola. The two plots are shown in Figure 4.12.

We observe that the graphs intersect only at the abscissa

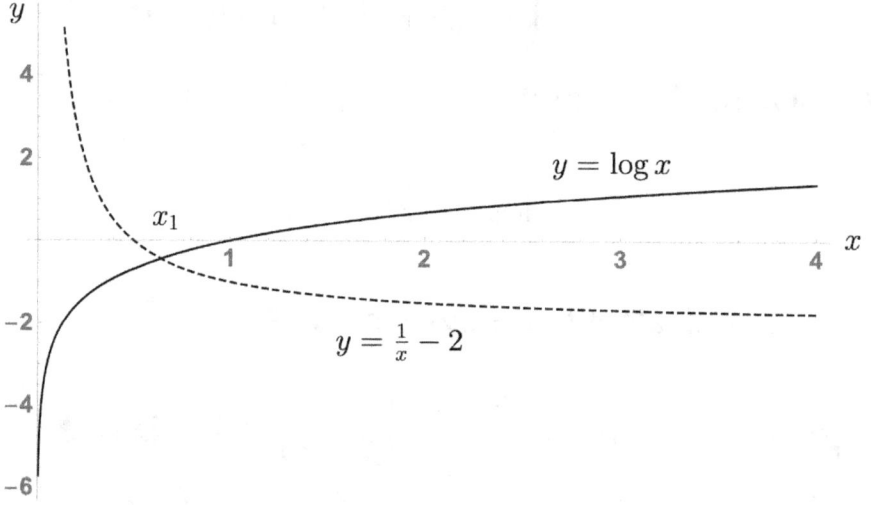

Figure 4.12: Plot of the functions $y = \log x$ (continuous line) and $y = 1/x - 2$ (dashed line).

point x_1 with $0 < x_1 < 1$. The inequality is satisfied for $0 < x < x_1$.

For the second inequality, $\log x > -1/x - 2$, we proceed in

4.22 Exercise 22

the same way, drawing the plots of the functions $y = \log x$ and $y = -1/x - 2$. The two are shown in Figure 4.13. We observe that the graphs never intersect, furthermore we

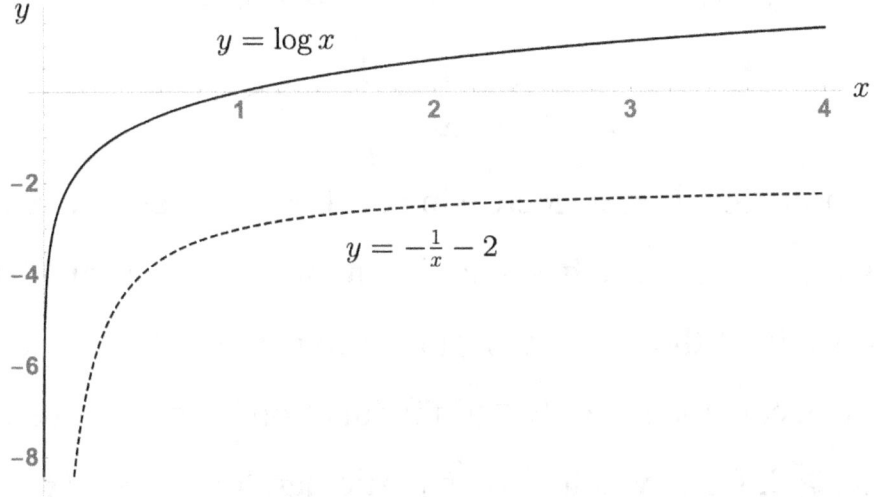

Figure 4.13: Plot of the functions $y = \log x$ (continuous line) and $y = -1/x - 2$ (dashed line).

deduce that $\log x > -1/x - 2$. We show this by considering the difference function defined by

$$f(x) = \log x + \frac{1}{x} + 2,$$

that must always be positive, i.e. $f(x) > 0$. The first derivative is

$$f'(x) = \frac{1}{x} - \frac{1}{x^2},$$

solving $f'(x) > 0$ we obtain, remembering that $x > 0$,

$$\frac{1}{x} - \frac{1}{x^2} > 0, \quad \frac{1}{x} > \frac{1}{x^2}, \quad x > 1.$$

The function $f(x)$ is decreasing for $0 < x < 1$ and increasing for $x > 1$, it also has a minimum in $x = 1$. We calculate the value of the minimum $f(1) = \log(0) + 1 + 2 = 3$ and we can conclude that $\forall x \in \mathcal{D}$ the function is positive, being $f(x) \geq 3$, from which also, in particular, $\log x > -1/x - 2$. The system relative to the inequality $|x(\log x + 2)| < 1$ can be simplified with the single inequality

$$0 < x < x_1.$$

The series is **absolutely convergent** and then also **convergent** if $x \in E$ with

$$E \equiv \{x \in \mathbb{R} : 0 < x < x_1\},$$

4.22 Exercise 22

absolutely divergent if $x \in F$ with

$$F \equiv \{x \in \mathbb{R} : x > x_1\},$$

with unknown behavior if

$$x \in S \equiv \{x_1\}.$$

We find the values of x for which the series is both absolutely divergent and with positive terms that can be obtained intersecting the respective sets A and F. To do this, it is necessary to define the hierarchy between e^{-2} and x_1. We consider the function

$$g(x) = -\log x + \frac{1}{x} - 2,$$

remembering that x_1 is the solution of the equation $g(x) = 0$ (see the first inequality of the system, whose solution was determined by considering the plots in Figure 4.12), i.e. $g(x_1) = 0$. We calculate the value that the function $g(x)$ assumes in e^{-2}, we have

$$g(e^{-2}) = -\log e^{-2} + \frac{1}{e^{-2}} - 2 = 2\log e + e^2 - 2 = e^2.$$

The function $g(x)$, for $x > 0$, has always a negative derivative

$$g'(x) = -\frac{1}{x} - \frac{1}{x^2} < 0,$$

and $g(x)$ is a decreasing function. So, being

$$g(x_1) = 0 < e^2 = g(e^{-2}),$$

we can conclude that $x_1 > e^{-2}$, hence also $e^{-2} < x_1 < 1$. Intersecting the sets A and F, we obtain

$$D \equiv A \cap F = \{x \in \mathbb{R} : x > x_1\}.$$

For $x \in D$ the series is **divergent**, because the study of absolute divergence coincides with that of simple divergence, being a positive terms series.

The series is never simultaneously with alternate sign terms and absolutely divergent, in fact, by intersecting the respective sets, we obtain the empty set

$$B \cap F = \{\}.$$

4.22 Exercise 22

We determine the behavior of the initial series for the values $x \in S$ for which the asymptotic ratio criterion has not provided informations, i.e., for $x = x_1$. In this case we have $x(\log x + 2) = 1$ and the series becomes

$$\sum_{n=1}^{\infty} \frac{1}{n \arctan n}.$$

To obtain its behavior we apply the asymptotic ratio criterion with the divergent harmonic series

$$\sum_{n=1}^{\infty} \frac{1}{n}.$$

We calculate

$$\lim_{n \to \infty} \frac{1/(n \arctan n)}{1/n} = \lim_{n \to \infty} \frac{n}{n \arctan n} = \lim_{n \to \infty} \frac{1}{\arctan n} = \frac{2}{\pi}$$

from which we conclude that the two series have the same behavior and therefore that for $x = x_1$, the initial series is **divergent**.

Summarizing:

The **SERIES** is:

- **CONVERGENT** if

$$0 < x < x_1;$$

- **DIVERGENT** if

$$x \geq x_1.$$

Summary scheme:

	0		x_1	
∄	∄	C	D	D

We can find an approximation of x_1 which represents the solution of the equation $x(\log x + 2) = 1$. We already know looking at the plot that $0 < x_1 < 1$, moreover, during the exercise we also found that $e^{-2} < x_1 < 1$. The quantity x_1 can be seen as a zero of the function $h(x) = x(\log x + 2) - 1$. The first derivative is $h'(x) = \log x + 2 + x(1/x) = \log x + 3$. We calculate $h'(x) > 0$, $\log x > -3$, $x > e^{-3}$. Since $e^{-3} < e^{-2}$ and we only consider the interval $e^{-2} < x < 1$, we can say that for x in this interval the function

4.22 Exercise 22

$h(x)$ is increasing. We calculate the value that the function $h(x)$ assumes at the ends of the interval, we have $h(e^{-2}) = e^{-2}(\log e^{-2} + 2) - 1 = -1 < 0$, $h(1) = \log 1 + 2 - 1 = 1 > 0$. We calculate its value at the middle point abscissa $(e^{-2}+1)/2$, we have $h((e^{-2}+1)/2) \simeq -0.186 < 0$, being negative as well as $h(e^{-2}) < 0$ we can conclude that $(e^{-2}+1)/2 < x_1 < 1$. Continuing with the bisection method we calculate $h((e^{-2}+3)/4) \simeq 0.377 > 0$. We deduce that $(e^{-2}+1)/2 < x_1 < (e^{-2}+3)/4$, we calculate $h((3e^{-2}+5)/8) \simeq 0.087 > 0$ (being $(3e^{-2}+5)/8$ the average value between $(e^{-2}+1)/2$ e $(e^{-2}+3)/4$), hence $(e^{-2}+1)/2 < x_1 < (3e^{-2}+5)/8$ that we write as $0.568 < x_1 < 0.676$. Continuing

$(0.568+0.676)/2 \simeq 0.622$, $f(0.622) \simeq -0.051 < 0$,

$0.622 < x_1 < 0.676$, $(0.622+0.676)/2 \simeq 0.649$,

$f(0.649) \simeq 0.017 > 0$, $0.622 < x_0 < 0.649$,

$(0.622+0.649)/2 \simeq 0.636$, $f(0.636) \simeq -0.016 < 0$,

$$0.636 < x_0 < 0.649.$$

You can proceed until you reach the desired approximation. A numerical calculation gives the approximate solution $x_1 \simeq 0.642$.

4.23 Exercise 23

Text

Study the behavior of the series of functions

$$\sum_{n=1}^{\infty} \left(\frac{n+x}{n}\right)^n \frac{1}{n^2}$$

Solution

There are no limitations on the values that the x variable can assume, so the domain is $\mathcal{D} = \mathbb{R}$.

We solve the inequality

$$\frac{n+x}{n} > 0, \quad 1 + \frac{x}{n} > 0, \quad -1 < \frac{x}{n}, \quad n > -x.$$

We observe that the series has positive terms from a certain n onwards, i.e. when $n > -x$. This means that there exists $\bar{n} \in \mathbb{N}$ such that for every $n > \bar{n}$ we have $x + n > 0$. We study the following positive term series

$$\sum_{n=\bar{n}}^{\infty} \left(\frac{n+x}{n}\right)^n \frac{1}{n^2},$$

in fact, the modification (or cancellation) of a finite number of terms does not modify the behavior of the series.

We apply directly the asymptotic ratio criterion

$$\left|\frac{a_{n+1}}{a_n}\right| = \left(\frac{n+1+x}{n+1}\right)^{n+1} \frac{1}{(n+1)^2} \cdot \left(\frac{n+x}{n}\right)^{-n} n^2,$$

from which, being,

$$\lim_{n\to\infty} \frac{n^2}{(n+1)^2} = 1,$$

$$\lim_{n\to\infty} \left|\frac{a_{n+1}}{a_n}\right| = \lim_{n\to\infty} \left(\left(1+\frac{x}{n+1}\right)^{n+1} \left(1+\frac{x}{n}\right)^{-n}\right) = 1.$$

The criterion does not provide additional indications. We proceed with the root criterion

$$\sqrt[n]{a_n} = \left(\left(\frac{n+x}{n}\right)^n \frac{1}{n^2}\right)^{1/n} = \left(\frac{n+x}{n}\right) \frac{1}{n^{2/n}} = \frac{n+x}{n^{2/n+1}},$$

we have

$$\lim_{n\to\infty} \sqrt[n]{a_n} = \lim_{n\to\infty} \frac{n+x}{n \cdot n^{2/n}} = \lim_{n\to\infty} \frac{n+x}{n \cdot e^{2\log(n)/n}} = 1,$$

so even this criterion does not provide indications. We consider the argument of the series as the sequence of functions

$$a_n = \left(\frac{n+x}{n}\right)^n \frac{1}{n^2}, \quad n > \bar{n}.$$

4.23 Exercise 23

We observe that the necessary condition for convergence

$$\lim_{n\to\infty} a_n = 0,$$

is always satisfied, in fact

$$\lim_{n\to\infty} \left(\frac{n+x}{n}\right)^n \frac{1}{n^2} = e^x \lim_{n\to\infty} \frac{1}{n^2} = 0.$$

We continue with the asymptotic ratio criterion with the convergent generalized harmonic series of term n-th $b_n = 1/n^2$ and calculate the limit of the ratio a_n/b_n

$$\lim_{n\to\infty} \left(\frac{n+x}{n}\right)^n \frac{1}{n^2} \cdot n^2 = \lim_{n\to\infty} \left(1+\frac{x}{n}\right)^n = e^x,$$

being $e^x \in \mathbb{R}$ we conclude that the two series have the same behavior and the initial series converges, $\forall x \in \mathcal{D}$.

Summarizing:

The **SERIES** is:

- **CONVERGENT** if

$$\forall x \in \mathbb{R}.$$

Summary scheme:

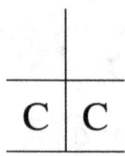

4.24 Exercise 24

Text

Study the behavior of the series of functions
$$\sum_{n=1}^{\infty} \left(\frac{x+3}{\sqrt{x}-2}\right)^n \frac{\log n}{n!}$$

Solution

As a condition of existence we have $x \geq 0$ due to the presence of the square root and $\sqrt{x} \neq 2$, i.e. $x \neq 4$. The domain is $\mathcal{D} = \{x \in \mathbb{R} : x \geq 0 \wedge x \neq 4\}$.

We solve the inequality
$$\frac{x+3}{\sqrt{x}-2} > 0.$$

The numerator is positive if $x+3 > 0$, i.e. for $x > -3$, while the denominator is positive if $\sqrt{x} - 2 > 0$, i.e. for $x > 4$. Combining these two results we obtain that the ratio is positive if $x > 4$, where we have also considered the domain.

The series is with **positive terms** if $x \in A$ with
$$A \equiv \{x \in \mathbb{R} : x > 4\},$$

with **alternate sign terms** if $x \in B$ with

$$B \equiv \{x \in \mathbb{R} : 0 \leq x < 4\}.$$

We consider the absolute series

$$\sum_{n=1}^{\infty} \left| \left(\frac{x+3}{\sqrt{x}-2} \right)^n \frac{\log n}{n!} \right| = \sum_{n=1}^{\infty} \left| \frac{x+3}{\sqrt{x}-2} \right|^n \frac{\log n}{n!},$$

in fact $\forall n \geq 1$ we have $\log n > 0$ and $n! > 0$. We apply the asymptotic ratio criterion

$$\left| \frac{a_{n+1}}{a_n} \right| = \left| \frac{x+3}{\sqrt{x}-2} \right|^{n+1} \frac{\log(n+1)}{(n+1)!} \cdot \left| \frac{\sqrt{x}-2}{x+3} \right|^n \frac{n!}{\log n}$$

$$= \left| \frac{x+3}{\sqrt{x}-2} \right| \cdot \frac{\log(n+1)}{(n+1) \log n},$$

$$\lim_{n \to \infty} \left| \frac{a_{n+1}}{a_n} \right| = \left| \frac{x+3}{\sqrt{x}-2} \right| \cdot \lim_{n \to \infty} \left(\frac{\log(n+1)}{(n+1) \log n} \right) = 0,$$

where we have used

$$\lim_{n \to \infty} \frac{\log(n+1)}{\log n} = 1.$$

The series is **absolutely convergent** and it is also **convergent** $\forall x \in \mathcal{D}$ and this concludes the exercise.

Summarizing:

The **SERIES** is:

4.24 Exercise 24

- **CONVERGENT** if

$$x \geq 0 \ \land \ x \neq 4.$$

Summary scheme:

	0		4	
$\nexists$	C	C	$\nexists$	C

4.25 Exercise 25

Text

Study the behavior of the series of functions

$$\sum_{n=2}^{\infty} \frac{n^{2n}}{n!} \left((x-5)(x+1)\right)^{n^2 \log n}$$

Solution

The base of the power must be a non-negative quantity, in fact the exponent, $n^2 \log n$, assumes non-integer values. To find the domain we need to solve the inequality

$$(x-5)(x+1) \geq 0.$$

The first factor is positive if $x-5 > 0$, i.e. for $x > 5$, the second factor is positive if $x+1 > 0$, i.e. for $x > -1$. Combining these results we obtain the following solution for the inequality

$$x < -1 \ \lor \ x > 5.$$

The domain is

$$\mathcal{D} = \{x \in \mathbb{R} : x \leq -1 \ \lor \ x \geq 5\}.$$

4.25 Exercise 25

For $x = -1$ and $x = 5$ the series has null terms and then it is simply **convergent**. For the other values that x can assume, the series has positive terms and therefore we can apply the asymptotic ratio criterion

$$\left| \frac{a_{n+1}}{a_n} \right| = \frac{(n+1)^{2(n+1)}}{(n+1)!} \left((x-5)(x+1) \right)^{(n+1)^2 \log(n+1)}$$
$$\cdot \frac{n!}{n^{2n}} \left((x-5)(x+1) \right)^{-n^2 \log n},$$

from which, being,

$$\frac{(n+1)^{2(n+1)}}{(n+1)n^{2n}} = \frac{(n+1)^{2n+1}}{n^{2n}} = \frac{(n+1)(n+1)^{2n}}{n^{2n}}$$
$$= (n+1)\left(1 + \frac{1}{n}\right)^{2n},$$

$$\left| \frac{a_{n+1}}{a_n} \right| = (n+1)\left(1 + \frac{1}{n}\right)^{2n}$$
$$\cdot \left((x-5)(x+1) \right)^{(n^2+2n+1)\log(n+1) - n^2 \log n},$$

and

$$\left|\frac{a_{n+1}}{a_n}\right| = (n+1)\left(1+\frac{1}{n}\right)^{2n}$$
$$\cdot \left((x-5)(x+1)\right)^{n^2\log(1+1/n)+(2n+1)\log(n+1)}$$
$$= (n+1)\left(1+\frac{1}{n}\right)^{2n}$$
$$\cdot \left((x-5)(x+1)\right)^{n\log(1+1/n)^n+(2n+1)\log(n+1)}.$$

Calculating the limit we obtain

$$\lim_{n\to\infty}\left|\frac{a_{n+1}}{a_n}\right| = \lim_{n\to\infty}(n+1)\left(1+\frac{1}{n}\right)^{2n}$$
$$\cdot \left((x-5)(x+1)\right)^{n\log(1+1/n)^n+(2n+1)\log(n+1)}$$
$$= \lim_{n\to\infty}(n+1)\left[\left(1+\frac{1}{n}\right)^n\right]^2$$
$$\cdot \left((x-5)(x+1)\right)^{n\log(1+1/n)^n+(2n+1)\log(n+1)}$$
$$= \begin{cases} 0 & \text{if } (x-5)(x+1) < 1 \\ \infty & \text{if } (x-5)(x+1) \geq 1 \end{cases},$$

4.25 Exercise 25

in fact in the case $(x-5)(x+1) = 1$ the limit is reduced to

$$\lim_{n \to \infty} \left| \frac{a_{n+1}}{a_n} \right| = e^2 \lim_{n \to \infty} (n+1) = \infty,$$

where we have used the notable limit

$$\lim_{n \to \infty} \left(1 + \frac{1}{n}\right)^{2n} = \lim_{n \to \infty} \left[\left(1 + \frac{1}{n}\right)^n\right]^2 = e^2.$$

We solve the inequality

$$(x-5)(x+1) < 1, \quad x^2 - 4x - 5 < 1, \quad x^2 - 4x - 6 < 0,$$

the associated equation has the solutions $x^2 - 4x - 6 = 0$, $\Delta = 16 + 24 = 40 = 10 \cdot 2^2$, $x_{1,2} = (4 \pm 2\sqrt{10})/2$, $x_{1,2} = 2 \pm \sqrt{10}$. The solution of the inequality can be written as

$$2 - \sqrt{10} < x < 2 + \sqrt{10},$$

combining with the domain $\mathcal{D}$ we have

$$2 - \sqrt{10} < x \leq -1 \ \lor \ 5 \leq x < 2 + \sqrt{10}.$$

The series is **convergent** if $x \in E$ with

$$E \equiv \{x \in \mathbb{R} : 2 - \sqrt{10} < x \leq -1 \lor 5 \leq x < 2 + \sqrt{10}\},$$

and **divergent** if $x \in F$ with

$$F \equiv \left\{ x \in \mathbb{R} : x < 2 - \sqrt{10} \lor x > 2 + \sqrt{10} \right\}.$$

Summarizing:

The **SERIES** is:

- **CONVERGENT** if

$$2 - \sqrt{10} < x \leq -1 \lor 5 \leq x < 2 + \sqrt{10};$$

- **DIVERGENT** if

$$x \leq 2 - \sqrt{10} \lor x \geq 2 + \sqrt{10}.$$

Summary scheme:

	$2-\sqrt{10}$		-1		5		$2+\sqrt{10}$	
D	D	C	C	∄	C	C	D	D

www.ingramcontent.com/pod-product-compliance
Lightning Source LLC
Chambersburg PA
CBHW080454220526
45465CB00006B/2264